The Low Cost Planet

The Low Cost Planet

Energy and Environmental Problems, Solutions and Costs

Dave Toke

Pluto **Press**

LONDON • BOULDER, COLORADO

First published 1995 by
Pluto Press
345 Archway Road
London N6 5AA
and 5500 Central Avenue
Boulder, Colorado 80301, USA

British Library Cataloguing in Publication Data
A catalogue record for this book is available from
the British Library

ISBN 0 7453 0843 0 hb

Library of Congress Cataloging in Publication Data
A catalog record for this book is available from
the Library of Congress

95   96   97   98   99   5   4   3   2   1

Designed and produced for Pluto Press by
Chase Production Services, Chipping Norton, OX7 5QR
Typeset from author's disk by
Stanford DTP Services, Milton Keynes, MK17 9JP

Contents

Tables

Figures

Acknowledgements

This book would not have been possible without the cooperation of a wide number of individuals, agencies and companies who have provided me with reports, data and comments. The individuals are too numerous to mention in full but include officials from the US Environmental Protection Agency (EPA), the Public Utility Commissions of various US states, Dutch and Danish energy companies, the UK's Energy Technology Support Unit (ETSU) and Warren Springs Laboratory.

Particular thanks go to Tim Woolf of the Tellus Institute in Boston, Eric Hirst of the Oakridge National Laboratory, Angie Minkin and Tom Thompson of the California Public Utilities Commission, and Bob Everett, Godfrey Boyle and David Olivier of the (UK) Open University's Energy and Environment Research Unit.

Special thanks also to those who offered comments on early drafts of some chapters, including Dave Elliot (Chapters 7, 11 and 12), Sue Walker and Mel Harvey (Chapter 5), John Thornes (Chapter 2) and Stephen Potter who helped me greatly with Chapter 8.

I also would like to thank Chris Cragg for explaining some of the vagaries of the oil industry, Jonathan Stern and Mike Prior for answering some questions about the gas industry, and Jackie Huang for some general comments.

Although this book does not contain a formal bibliography, the references made in the Notes are intended as a useful guide to sources. The Notes also provide additional elucidations of points made in the main text.

International energy statistics quoted in this book are derived from the BP *Statistical Review of World Energy*, except where otherwise stated.

References to *The Economist* relate to the UK edition. References to *Energy Economist* relate to the *Financial Times* newsletter of that name.

The illustrations were drawn by Richard Ellis.

Glossary

The various costs quoted throughout this book are in 1993 prices, except where otherwise stated. The costs are in either UK sterling or US dollars, depending on the source of the data. In 1993 there were roughly $1.50 to the pound, so that, for example, 3 pence per kWh (3p/kWh) was equivalent to 4.5 cents/kWh and $3 per GJ was equivalent to £2 per GJ.

Many terms are explained in the text. Some basic and recurring terms are as follows:

W	watt; a unit of power of one joule per second
kW	1,000 watts
MW	1 million watts and 1,000 kilowatts
GW	1 million kilowatts

The terms kW, MW, GW can describe power in the form of heat or electricity. Electrical power is often distinguished by adding an 'e' (for example, MWe). However, in this book it should be assumed that the terms kW, MW and GW refer to electrical power, unless otherwise stated.

EPA	Environmental Protection Agency, a regulatory body set up under the aegis of the US Federal Government
EU	European Union
GDP	gross domestic product, a measure of a nation's output of goods and services
GJ	gigajoule; there are roughly 278 kWh in a GJ
GT	gigatonne or 1,000 million tonnes (10^9 tonnes)
IEA	International Energy Agency (this consists mainly of developed countries and was formed as a response to the 1970s' oil crisis)
km	kilometre
kph	kilometres per hour
kWh	kilowatt hour; the energy transferred by a power of 1 kW in an hour
MTOE	million tonnes of oil equivalent
OECD	Organisation for Economic Cooperation and Development (mainly 'developed' countries)

1

Introduction

Nothing moves without energy, and no energy can be used without disturbing the environment.

As the average motorist drives down the road he or she may (sometimes) feel vaguely guilty about the bluish-black smog that envelops the metropolis he is approaching, but he will usually be pretty bewildered about what measures are required to solve such environmental problems.

This book, therefore, sets out to describe our main energy–environmental problems, to examine solutions to these problems and to compare the costs of the different solutions. In doing this I shall assess the accuracy of the established view which says that the solution of fundamental environmental problems will increase the monetary costs of supplying energy services.

Patterns of energy consumption emerge as a result of a complex interaction of the forces that shape energy markets. The road to the energy future is paved by technology, but the direction which this road takes is determined by culture and the aspirations of the dominant energy interest groups. These groups influence governmental policies, and by their own activity they determine the nature of the market.

It is important to spell out the nature of these 'players' in the energy game from the start.

First, there are the key energy production companies. These include the main oil companies such as Shell and Exxon. The domestic generators and distributors of energy products are also crucial parts of the energy supply industries.

Second, there are the energy supply countries. These have grown more important since the emergence of the Organisation of Petroleum Exporting Countries (OPEC). Some countries will have more cheaply available resources and cheaper labour costs than others.

Third, there are the countries that demand energy. Unless nations are energy self-sufficient, energy imports can impose considerable balance of payments problems and depress the value of their currencies. Even the US, the country which gave birth to the oil industry, now fears the impact of oil price rises. Other, poorer, developing nations can be devastated by energy crises.

A crucial fourth group is comprised of the individual final energy consumers. They want reliable, high quality, convenient energy services at low cost. Traditionally, this demand has been interpreted as being for energy. However, consumers want buildings that are well lit, televisions that work and so on, not energy per se. Though theoretically immensely powerful, energy consumers are fragmented into various types of industrial, commercial and domestic groups which are difficult to organise.

These four groups have been around for a very long time. However, it has been only since the 1960s that we have seen the emergence of an influential and identifiably 'green' lobby whose objectives have been to combat pollution and to protect natural habitats. They have, in recent years, gained a great deal of publicity, but this should not be confused with political strength.

It has traditionally been the custom to separate 'environmental' targets from other objectives. It is convenient for established energy supply groups to do this, but the crucial question is whether the attainment of environmental objectives can help, rather than obstruct, the attainment of conventional objectives as expressed by the interest groups themselves. Indeed, it is often difficult to separate the conventional from the environmental concerns. For example, resource depletion and energy dependence are both conservationist and conventional concerns.

This book is international in scope. Many issues and problems are common to both developed and developing nations. Specific attention is given, where appropriate, to issues that affect developing countries and Eastern Europe, which shares some of the problems faced by developing nations. I use the term 'Eastern Europe' to refer to those European countries (including the eastern part of Germany) which used to be part of the communist bloc.

The Low Cost Planet has the following structure. Chapter 2, on pollution problems, describes the causes and consequences of acid rain, smog, contributions to so-called global warming, oil spills and electromagnetic fields. International efforts to control and reduce these problems are considered. Resource problems, again affecting fossil fuels (coal, oil and natural gas), are discussed in Chapter 3. Chapter 4 is a short account of some of the main types of solutions and the strands of thought in the energy debate.

It is essential that some common standard is established for assessing the costs of different energy options. This is the focus of Chapter 5. Chapter 6 looks at the use of natural gas as a way of dealing with environmental and resource problems. Chapter 7 looks at the possibilities for energy efficiency, the technology involved in its implementation and the policy strategies that can be employed to make it happen. The issue of combating pollution from motor cars is the subject of Chapter 8.

Chapter 9 looks at the techniques of reducing the environmental consequences of coal use. Chapter 10 investigates means of removing and disposing of carbon dioxide emitted by fossil fuel burning.

Attention is then turned to non-fossil fuels. First, nuclear power: Chapter 11 looks at its potential, costs and environmental impacts. Then, in Chapter 12, renewable energy. This is in fact an umbrella term for a range of fuels that often differ as much between themselves as they do with fossil fuels.

Finally, Chapter 13 draws some conclusions together.

2

The Pollution Problem

The world's energy economy is dominated by fossil fuels. They comprise nearly 90 per cent of officially recorded energy supplies. They are also held responsible for many of the world's worst environmental problems.

This chapter concentrates on fossil-fuel-related pollution problems. The environmental impact of alternative, non-fossil, energy sources will be examined in later chapters.

Acid Rain, Smog and Pollution from Cars

Acid rain is a vague term, which is not surprising since it dates from the nineteenth century, long before there was much scientific analysis of air pollution problems. I define acid rain as the deposition of sulphur dioxide (SO_2) and oxides of nitrogen (NOx).[1]

Some of these emissions are actually deposited in a dry form consisting of gases and particles, usually quite near the source. The rest of the gases are dissolved in water to form acid droplets which are deposited sometimes hundreds or even thousands of miles away from the pollution source.

The Extent and Trends of Acid Deposition

Sulphur dioxide (SO_2) and nitrogen oxides (NOx) are produced by power stations, industry and motor cars.

Figures 2.1 to 2.4 show the origins of sulphur and nitrogen emissions in the US and the UK. Sulphur comes mostly from power stations (mainly coal-fired) and industry (especially from fuel used in metal smelting). The NOx come principally from motor vehicles although power stations and industry play a large role as well.

Figures for acid emissions from different countries are given in Table 2.1 (SO_2) and Table 2.2 (NOx).

It can be seen that the US and Canada are very big producers of acid material, although in per capita terms the former East Germany beat everyone in 1990. West Germany is relatively low, and Japan is especially low because of the acid emissions reduction programmes effected in those countries.

4

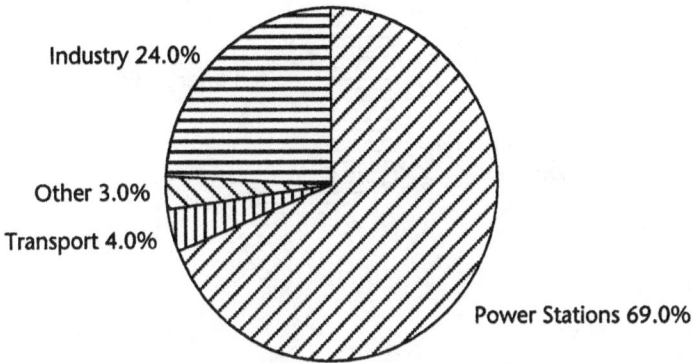

Figure 2.1 US SO$_2$ Sources (Per Cent Contributions), 1989

Source: EPA, *EPA's Acid Rain Program*, Washington DC, 1991

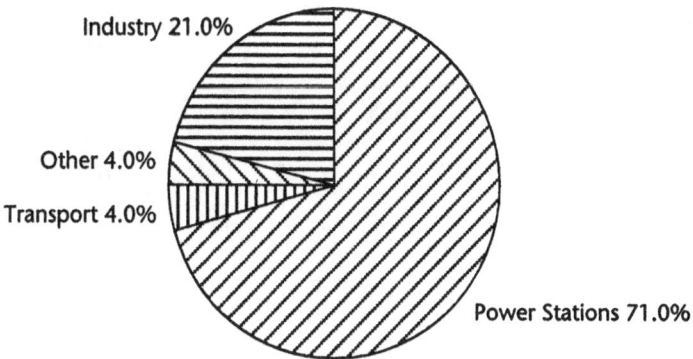

Figure 2.2 UK SO$_2$ Sources (Per Cent Contributions), 1992

Source: DOE, *Environmental Data*, London, 1993

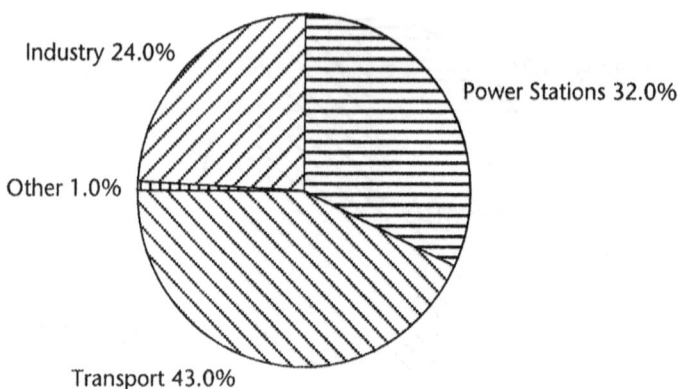

Figure 2.3 US NOx Sources (Per Cent Contributions), 1989

Source: EPA, *EPA's Acid Rain Program*, Washington DC, 1991

Table 2.1 Past SO_2 Emissions in Selected Countries (1,000s Tonnes)

	1970	1980	1991
Canada*	6,677	4,643	3,306
USA*	28,420	23,780	20,730
Japan*	4,973	1,263	876[1]
Denmark*	574	449	181[2]
France*	2,966	3,348	1,314
West Germany*	3,743	3,194	939[2]
East Germany	4,114[3]	4,323	4,758[2]
Hungary	–	1,663	1,085[1]
Netherlands*	807	502	204
Poland	–	4,100	2,995
Sweden*	930	489	106
Turkey*	–	–	1,602[1]
Russian Federation	–	7,161	4,460[2]
Ukraine	–	–	2,196[2]
UK*	6,424	4,898	3,780
OECD (Total)	64,900	53,900	40,200[2]

1. 1989
2. 1990
3. 1975
*indicates OECD membership
Dashes indicate that insufficient data are available
Sources: OECD, *Digest of Environmental Data* (Paris, 1993); UN European Commission for Europe

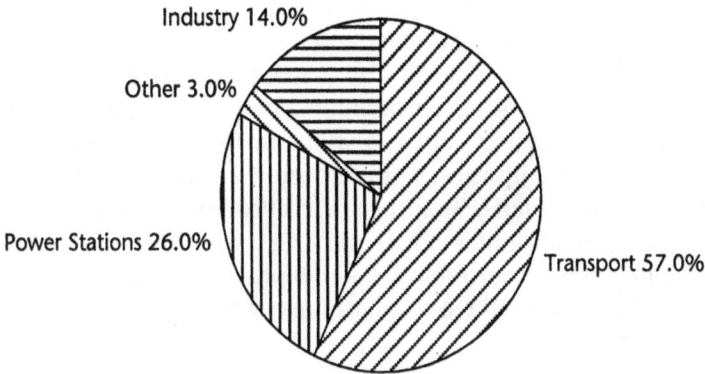

Figure 2.4 UK NOx Sources (Per Cent Contributions), 1992

Source: DOE, Environmental Data, London, 1993

Table 2.2 Past NOx Emissions in Selected Countries (1,000s Tonnes)

	1970	1980	1991
Canada	1,364	1,959	1,923[1]
USA	18,960	23,560	18,760
Japan	1,651	1,400	1,301[2]
Denmark	–	270	283[1]
France	1,322	1,646	1,507
West Germany	2,345	2,944	2,605[1]
East Germany	–	593	629[1]
Hungary	–	273	249[2]
Netherlands	456	571	550
Poland	–	–	1,205
Sweden	302	424	389
Turkey	–	380	175[2]
Ukraine	–	–	1,097[1]
UK	2,293	2,265	2,779
OECD (Total)	32,900	40,700	36,700[1]

1. 1990
2. 1989
Sources: OECD; UN European Commission for Europe

Acid Lakes and Rivers

Acidity is measured by pH values on a range between 0 and 14, the lower values being acidic and 7 being neutral. A pH of 5 is generally agreed as being the lower limit for the long-term survival of most types of fish and other aquatic animals. Trout and salmon tend to be especially sensitive to acid and are severely affected at pH values lower than 6. Toxic metals leached from rocks and soils by the acid attack gills and fish eggs.

As can be seen from Table 2.3 a high proportion of lakes have pHs of less than 5 and recovery is, in many places, proving to be a slow process.

Table 2.3 Acidity of Lakes (pH)

	1980	1991
Canada (Great Lakes)	4.4	4.32
New Zealand (Wellington)	5.12	5.14
Netherlands	4.46	4.70[1]
Norway (Birkenes)	4.16	4.37[1]
Poland (Suwalki)	4.45	4.27[1]
Slovak Republic (Bratislava)	4.18	4.66
UK (Inverpolly)	4.7	4.82

1. 1990
Source: OECD, *Digest of Environmental Data* (Paris, 1993)

The pH of lakes in Sweden and Norway was above 6 before the Second World War, but the pH of many lakes was, in the 1970s, well below 5. Local outrage at acid rain damage is particularly widespread due to the Scandinavian pastime of summer fishing holidays around northern rivers and lakes, a practice which has been constrained by the effects of acid rain.

In eastern Canada around 14,000 lakes are described as acidic by the Canadian government.

In the UK Scottish lakes are, as in Scandinavia, often easily acidified because of the igneous rocks which hold them. Lakes in areas dominated by sedimentary rock formations often fare better because limestone forms a carbonate 'buffer' solution which neutralises some of the acidity.

Tree Damage

The image of stumps of dead, allegedly acid-struck trees was perhaps the most powerful symbol of ecological peril that transformed the West German Die Grünen (Green Party) into a major political force in the early 1980s. This heralded international agreement to reduce sulphur emissions in Europe.

Paradoxically, acidification has never been conclusively proved to have been the sole major cause of German tree death. Ozone (described later under 'photochemical smog') is thought by many to be a major contributor. There have been forest losses throughout West Germany and Northern Europe on over a million hectares of land. In Eastern Europe trees in areas such as the so-called 'black triangle' – in northwest Bohemia in the Czech Republic near the German border – are said to be subject to very high levels of acid attack.[2]

In the UK and the US concern about acid rain lagged behind that shown by some other Western states in the 1970s and 1980s. Yet it has been estimated that around two-fifths of all British trees have suffered moderate to severe defoliation.[3] Forest die-back among red spruce in the northeastern US has been documented since the 1960s, and spruce in the high ridges of the Appalachians as well as the Shenandoah and Great Smoky National Parks have been well documented. The Great Smoky has become distinctly smoggy.

The major causes of tree decline are thought to be the washing away of nutrients and the ingestion of heavy metals released by the acids. Coniferous trees tend to be affected more severely by acid damage than hardwoods.

Crops can be damaged by both wet and dry deposition, although as with trees it is not often clear to what extent ozone and acids are respectively to blame.

Effects on Buildings and Infrastructure

Limestone buildings and metal structures have been among the parts of our infrastructure most severely affected by acid rain. In Upper Silesia, Poland, railway tracks have been gobbled up by acid to such an extent that the trains have been limited to 40 kph.[4]

Famous buildings ranging from the Parthenon in Athens to the Taj Mahal in the Indian city of Agra have been badly scarred by the effects of air pollution in recent decades, although the extent to which people's lungs have been eaten away has received less publicity.

Effects on Human Health

If lakes and rivers are acidified, then this is likely to have an effect on water supplies for humans, though there is uncertainty about the health effects. Some say that aluminium being leached into the water supply by acid causes Alzheimer's disease. High levels of lead can result from acidic water running through lead pipes, and lead poisoning can result in soft-water areas.

SO_2 and NOx emissions can have serious consequences for human health if they occur in high concentrations in the air that we breathe. The effects can be especially serious if these substances are ingested along with other airborne pollutants that are produced by the combustion of fossil fuels. Dust, whose technical term is 'particulates', can carry carcinogenic heavy metals like mercury that originate from coal.

A district in the Katowice area of Poland, for example, receives 1 kg of dirt per metre per year and the cancer rate is 30 per cent higher than in the rest of Poland, which is not exactly sparklingly clean.[5] High rates of respiratory diseases and allergic responses have been recorded in many areas of the former Eastern Europe as well as in many cities in developing countries that have high levels of sulphur and dust emissions. These emissions come mainly from coal-burning in power stations and industry. Some data on emissions in some mainly Eastern cities are given in Table 2.4.

In the past the horrors of such pollution were also visited on the West which used to be regularly subjected to bouts of debilitating fogs. These fogs, called 'pea soupers' in the UK, resulted in calamitous death tolls. Some 4,000 people are said to have died in the notorious London 'pea souper' of 1952.

In the UK, Clean Air legislation (from 1956 onwards) banned black smoke production in certain areas. Later on the use of tall chimney stacks, which sent the pollution elsewhere, was mandated. However, a switch in fuels from coal to natural gas (low in both dust and sulphur) for domestic cooking and heating was probably the most effective means of clearing the air.

Although SO_2 and dust pollution are much less of a problem in most parts of the West than they once were, NOx emissions, especially from motor cars, are still a major problem. NOx emissions are suspected of contributing to asthma and other respiratory illnesses.

Table 2.4 Air Pollutant Concentrations in Major Cities of Developing Countries

		SO$_2$ (no. of days over 150 µg/m$^{3)}$	Particulates (no. of days over 230 µg/m^3)
Brazil	Rio de Janeiro	–	11
	São Paulo	12	–
China	Beijing	68	272
	Shanghai	16	133
India	Calcutta	25	268
	Delhi	–	294
Indonesia	Jakarta	–	173
Iran	Teheran	104	174
Malaysia	Kuala Lumpur	0	37
Philippines	Manila	24	14
Korea	Seoul	87	–
Thailand	Bangkok	0	97
Japan	Tokyo	0	2

Note: 150 µg/m^3 of SO$_2$ and 230 µg/m^3 are former guideline values of the World Health Organisation (WHO). These guideline values have since been reduced
Source: GEMS, 'World Resources 1990–91', cited by Environment Agency of Japan

International Efforts to Control Acid Emissions

Acid rain takes no heed of international borders and so international action has been organised to tackle the problem. Negotiations about limiting acid emissions began in earnest after the Helsinki Conference in 1975. The negotiations culminated in the 1979 Geneva Convention on Long Range Transportation of Air Pollution. This Convention was strengthened over the next few years and progress was monitored by the United Nations Economic Commission for Europe (UNECE).

The Japanese and the Germans took the lead in establishing acid rain abatement programmes. The Anglo-Saxons in the US, Canada and the UK were more laggardly. The UK was typecast as the 'dirty man of Europe' by Scandinavians in particular who resented unwanted imports of British acid emissions.

However, the UK was forced to go along with European Community plans to cut acid emissions from power stations under the Large Combustion Plant Directive, agreed in 1988. This required the richest EC states to achieve 60–70 per cent reductions in 1980 levels of SO_2 emissions and 40 per cent (UK, 30 per cent) reductions in 1980 levels of NOx emissions by the year 2003.

The British strategy towards achieving its acid emission reduction targets consists of switching from coal to natural-gas-fired power stations, importing low sulphur coal, and fitting a minority of the existing coal-fired power stations with equipment that removes SO_2 emissions. Low NOx burners are also being fitted.

Japan and Germany began cleaning their coal-fired power stations in the 1970s and 1980s, so their SO_2 emissions are relatively low. The French have reduced their acid emissions through the nuclear power station programme.

A new Geneva Convention was signed, under UNECE auspices, in 1994, but the agreement fell short of UNECE objectives. Although some countries agreed to speed up their SO_2 emission reduction programmes, some nations, including the UK, refused to commit themselves to much more rapid progress. The UK agreed to reduce 1980 levels of SO_2 emissions by 50 per cent by 2000, 70 per cent by 2005 and 80 per cent by 2010.

Successive US governments did little to respond to demands for action on acid emissions until 1989 when George Bush agreed to introduce what became the 1990 Clean Air Act. The following year he signed an Air Quality Agreement with Canada. This promised a reduction of 40 per cent of 1980 US SO_2 emissions by the year 2010. A reduction of around 30 per cent of 1980 levels of Canadian SO_2 emissions is promised by the same date.[6]

The first phase of the US programme, which deals with the big coal-fired stations in the northeast and Mid-West, takes effect in 1995. The second, stricter phase, begins in the year 2000. As in the UK, the US path towards acid emission reductions is likely to consist of a mixture of switching to low sulphur coal, retrofitting existing coal-fired power stations and switching to natural gas power stations.

The US SO_2 abatement programme has divided environmentalists because of the use of the so-called 'emissions trading scheme'. The Environmental Protection Agency (EPA) administers the scheme under which power stations can emit pollution above a certain level only if they buy 'right to pollute' permits. The permits can be traded on the open market. The EPA controls the amount of pollution through the issuing of permits. The scheme was designed by the Environmental Defense Fund (EDF) but is opposed by some key environmental groups such as the Sierra Club. They object to the idea that people can buy the right to pollute, claiming that SO_2 emissions may become concentrated in some areas, subjecting them to serious damage.

Supporters of this market-based incentives scheme say that it will reduce the costs of implementing the emissions targets (in comparison with 'inflexible' regulations adopted in Europe and Japan), since it allows companies to make the most cost-effective decisions. For example, a company can buy permits to continue polluting if it finds it is cheaper to pay someone else to reduce emissions by a given amount than it is to reduce its own emissions by the same quantity.

The tradeable permits system is being extended at a local level, starting with southern California, to cover NOx emissions as well.

Targets for reductions in US and Canadian NOx emissions seem to be fairly modest. The US has set a formal target of around a 10 per cent reduction of 1980 levels of NOx emissions by the year 2000, and the Canadians have not agreed to any reduction in NOx emissions at all.

The adoption of acid rain abatement targets has not abolished the problem of energy-related local air pollution in the West. This is because of the rise of the motor car. Emissions from motor cars are largely responsible for a phenomenon called photochemical smog, which, combined with traces of the more traditional 'dust and sulphur' pollutants, causes serious problems in many areas.

Photochemical Smog

Photochemical smog is formed by reactions involving hydrocarbons and nitric oxide emissions from motor vehicles and other sources. This occurs in high temperatures and long periods of sunshine containing ultra violet radiation. Ozone and PAN (peroxyacetyl nitrate) come into being as a result of these reactions. NOx, ozone, PAN and dust particles form a dirty blue haze. Ozone can travel for over 1,000 kilometres.

Photochemical smog causes eye, nose and throat irritation and respiratory problems. People who suffer from allergic responses to airborne stimuli are often sufferers from asthma, and their condition may be aggravated by this type of smog. Mexico City (which harbours a mixture of old-style industrial and photochemical smog), Athens, many cities in developing countries, Los Angeles and the surrounding South California basin, all have chronic difficulties with photochemical smog.

Ozone has been blamed for much forest damage, with the Ponderosa Pines of San Bernadino being among the more celebrated victims. Widespread crop damage has been reported.

Other Pollutants from Motor Vehicles

Carbon monoxide, which at some of the higher concentrations present on the urban roadside can cause headaches and perhaps contribute to long-term cardiovascular damage, is also emitted by cars. Bus drivers and taxi drivers will be affected more than most others. Motor vehicles also produce unburned hydrocarbons. These include polynuclear aromatic hydrocarbons (PAH), such as benzene, which are carcinogens.

Some researchers have suggested that fine particles emitted by motor cars called PM10 (they are defined as less than 10 microns across) may be causing large numbers of deaths associated with lung-related illnesses. It has been hypothesised that the small particles act to 'spear' chemicals like ozone and NOx deep into lung tissue, so amplifying their effects.

A study of several cities, conducted by the Harvard School of Health, found that fatalities from cardiopulmonary disease and lung cancer were 37 per cent higher in the most PM10-polluted city, Steubenville, Ohio, than in the least affected, Portage in Wisconsin. PM10 may be killing 10,000 people a year in England and Wales. There is likely to be an increased effort to control emissions of these particles.[7]

Controls over Motor Vehicle Emissions

Controls over a widening range of motor vehicle emissions have been tightened in successive US Clean Air Acts since the first in 1967. The 1990 Clean Air Act ordered a range of new measures and standards. Many areas of the US still suffer smog episodes (as measured by ozone levels) as can be seen in Figure 2.5. The ozone non-attainment areas include Philadelphia, Chicago, Houston, New York State and, quite clearly the worst, the Los Angeles Basin.

The situation in many European cities is no better, although it was only in 1992 that European Community motor vehicle emission standards were raised to levels comparable with those of the US. (The means of reducing the impact of motor vehicle emissions is investigated in Chapter 8.)

Figure 2.5 Ozone Non–attainment Areas in the US

Source: American Gas Association, *The Gas Energy Demand and Supply Outlook*, 1993, p. 19

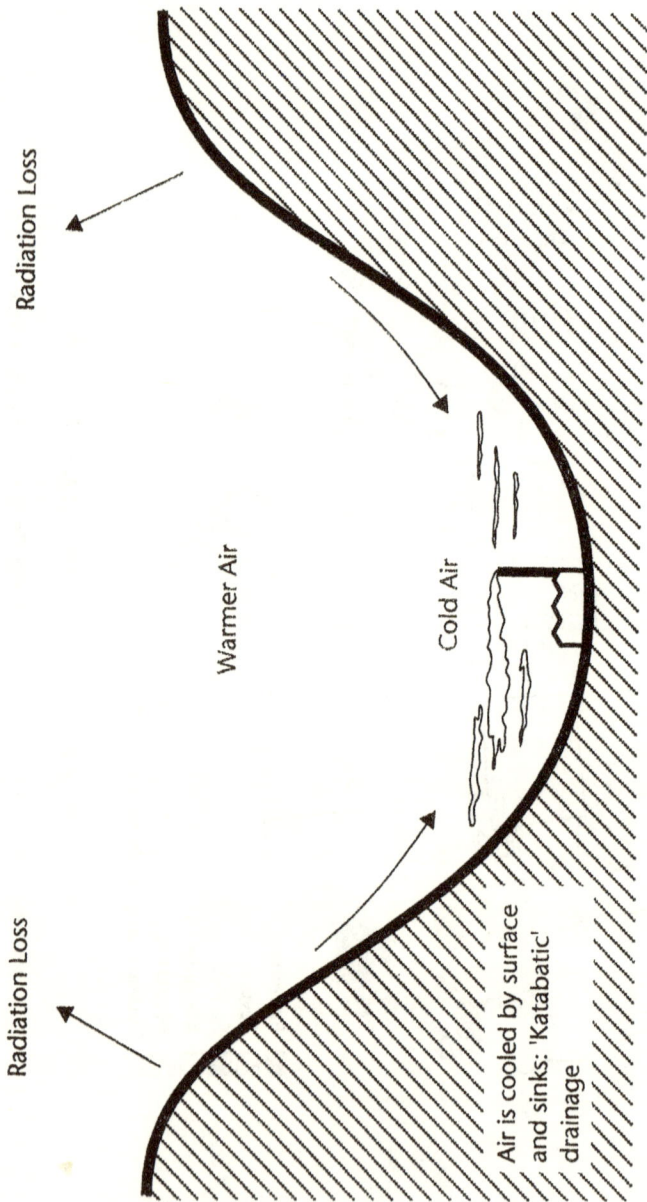

Radiation Loss

Radiation Loss

Warmer Air

Cold Air

Air is cooled by surface and sinks: 'Katabatic' drainage

Figure 2.6 Temperature Inversion

Temperature Inversions

Pollutants of various sorts are often such a severe problem in urban areas like Los Angeles and Athens because they are situated in valleys or river plains. This makes them susceptible to 'temperature inversions'.

Temperature inversions occur when there is high air pressure with little or no clouds. The slopes of a valley radiate at night causing adjacent air to be chilled. This air moves down the slopes as a cold wind, making the bottom of the valley cold. Thus normal temperature patterns are reversed and the layers nearest the floor of the valley are cold, with warm air above.

Pollution produced in the valley cannot rise as it usually does. It stays there until weather conditions change.

Air pollution, of one sort or another, has been a complaint for centuries. However, concern about the invisible effect of carbon dioxide emissions has a much shorter pedigree.

Global Warming

Concern that carbon dioxide emissions may significantly increase the temperature of the Earth's atmosphere dates back to the writings of the Swedish chemist Arrhenius in the 1890s. In the 1960s scientists like Bert Bolin started to flag the problem in more quantitative terms, although the issue remained relatively obscure until the 1980s.

Global environmental concerns shot up the political agenda in 1985 following the publication of measurements which suggested that stratospheric ozone over the Antarctic was being severely depleted by chlorofluorocarbons (CFCs). Besides depleting ozone, CFCs are thought to be powerful agents of global warming. In 1987 the UN's Brundtland Report[8] on environment and development identified global warming as a major issue.

It was the political fall-out surrounding the US 1988 Mid-West drought that made the subject hit the headlines. Jim Hansen of NASA told a US Senate committee that he was '98 per cent certain' that global warming was the cause of the Mid-West drought. In June 1988, a conference of scientists in Toronto called for a 20 per cent reduction in 1987 global levels of carbon dioxide emissions. They called for this target to be achieved by the year 2005.

Within a year of the Toronto conference, television viewers throughout the industrialised world were being bombarded with a series of eco-disaster warnings concerning the possibilities of drought and drowning. However, boredom eventually set in and people became concerned with other issues, notably the confrontation with Iraq. This was followed by economic recession. Some began to question whether the threat was as real or as potent as claimed and asked whether the potential consequences of rising carbon dioxide levels were really so terrible. Let us look at the issue in detail.

The Theory of Global Warming

Without the mediating impact of the atmosphere the temperature at the Earth's surface would be like that of the moon. The mean annual temperature of the Earth would be 33 degrees Centrigrade lower than it is and there would be much wider day-to-night variations in temperature.

Incoming solar radiation hits the Earth and is re-radiated at longer, infra-red wavelengths. However, the so-called 'greenhouse gases' absorb this radiation, thus stopping or slowing down its return to deep space.

Most of this 'greenhouse effect' is caused by water vapour, but a small amount is caused by carbon dioxide.

The problem is that concentrations of carbon dioxide and other greenhouse gases have been increasing at an ever more rapid rate since the start of the industrial revolution. These gases include carbon dioxide, chlorofluoro- and other halo-carbons, methane and nitrous oxide. Their relative new contributions to alleged increases in the greenhouse effect are given in Figure 2.7.

Figure 2.8 shows the range of estimates for near term (man-made) global temperature changes prepared by the Intergovernmental Panel on Climate Change (IPCC). This body is sponsored by the United Nations Environment Programme (UNEP) and the World Meteorological Organisation (WMO) and, as such, propounds the 'establishment' view.

The upper estimates reflect what would happen, according to the projections, if historical trends of increasing carbon dioxide emissions continued (3.5 degrees increase over 1990 by 2100). The lowest scenario estimate assumes that carbon dioxide emissions and other greenhouse gas emissions will increase more slowly before declining by the end of the twenty-first century (1.5 degrees increase by 2100). As can be guessed from the range, the amount of future warming depends very much on how human activities change over the next century.

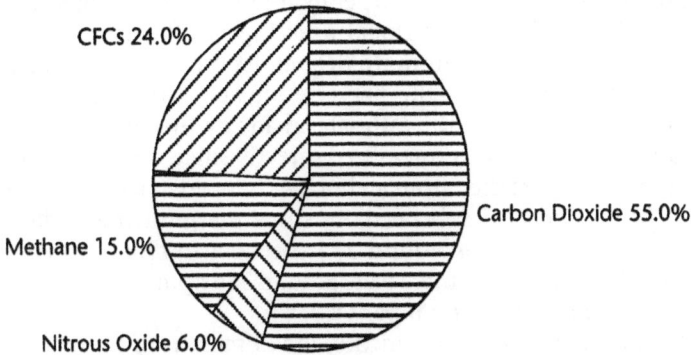

Figure 2.7 Relative Contribution of Greenhouse Gases to Radiative Change during the 1980s

Source: Mike Grubb, *Energy Policies and the Greenhouse, Effect* Royal Institute of International Affairs (Aldershot: Dartmouth Publishing, 1990), p. 14

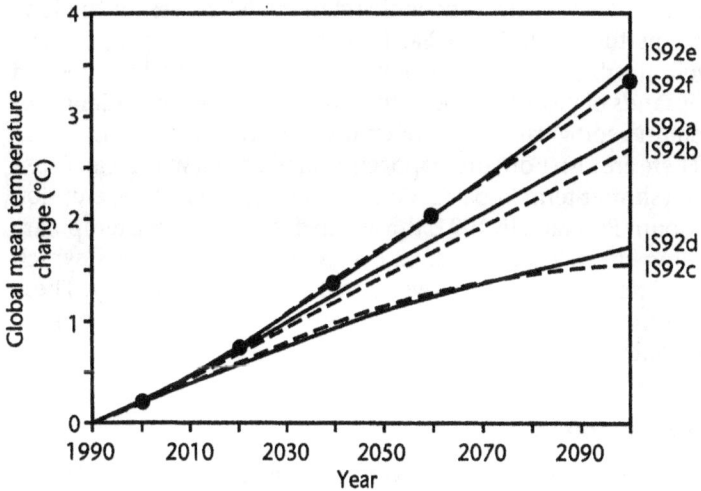

Figure 2.8 Mean Global Temperature Increase Projections (According to IPCC)

Note: IS92a to IS92f represent the various scenarios of different emission levels in the 1990 to 2100 period

Source: UNEP/WMO IPCC, B. Callender, J. Houghton, and S. Varney (eds.), *Climate Change 1992, Supplementary Report to the IPCC Scientific Assessment* (Cambridge: CUP, 1992), p. 18

A reduction in carbon dioxide emissions will (according to the theory) limit the extent of global warming. On the other hand, a successful drive to eliminate SO_2 emissions (which have a cooling effect) will have the perverse impact of increasing the warming effect beyond the estimates made by the IPCC.

Carbon dioxide levels (measured directly since the 1950s and extrapolated before then from air bubbles trapped in ice of known ages) have increased substantially since pre-industrial times. Before 1750 there were around 275 parts per million (ppm) of carbon dioxide. In 1992 there were around 356 ppm. The increase is currently around 1.5 ppm per year.

Six-sevenths of the carbon dioxide comes from fossil fuel burning, and most of the rest from deforestation. Taking account of the fact that up to a fifth of the methane and a small part of nitrous oxide is released as a result of fossil fuel use, energy use accounts for around half of the alleged new warming effect. CFCs account for another quarter and agricultural activities and waste disposal account for much of the rest.[9]

Of course, the Earth's temperature varies according to natural factors as well as anthropogenic (man-made) causes. In the long term, perturbations in the Earth's orbit affect global temperature. Cyclical orbital changes which last tens and even hundreds of thousands of years have been hypothesised. There is evidence for them in geological records of changes in ice-sheet cover.

There are also non-anthropogenic phenomena which can cause much shorter-term impacts. Volcanic activity, such as the eruption of Mount Pinutabo in 1991, throws up dust which has a temporary cooling effect. There are 'solar cycles', lasting a few years, which are defined as the time between peaks of sunspot activity. These cycles may involve changes in the amount of radiation from the sun (called 'solar irradiance').

Warming Impacts

The main impacts of global warming are likely to be on the sea level and also on rainfall patterns.

Sea level rises will result from thermal expansion of water and from melting of land-based ice-sheets. Using IPCC projections of warming, Wigley and Raper[10] say that sea levels will rise by around 55 cm by 2100 under the higher scenario (e) and around 35 cm under the lower scenario (c). It should be borne in mind that the 'lower' scenario assumes carbon dioxide emissions of only around 20 per cent below 1990 levels.

According to UN figures, global sea levels have risen by around 11 cm over the last 60 years.[11]

Figure 2.9 shows Wigley and Raper's projections of sea level rises.

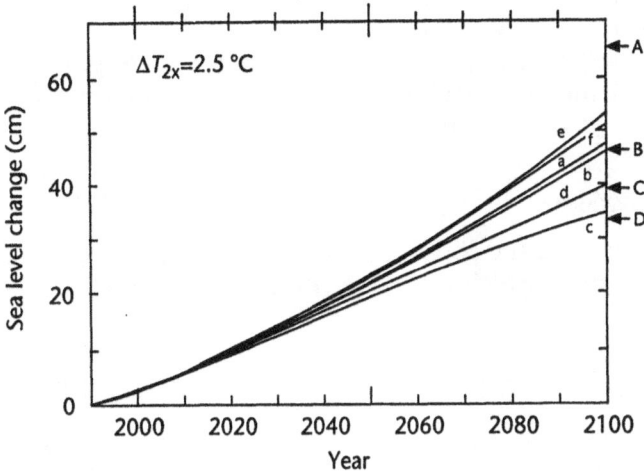

Figure 2.9 Projected Sea Level Rises Resulting from the Different IPCC Scenarios

Note: the right-hand indicators A, B, C, D represent results for the emissions scenarios published by the IPCC in 1990

Source: T.M.L. Wigley and S.C.B. Raper, 'Implications for Climate and Sea Level of Revised IPCC Emissions Scenarios', *Nature*, Vol. 357, No. 6376, May 1992

Although such sea level rises are not going to inundate major portions of land areas in themselves, they will make some of the smaller islands uninhabitable. Moreover, they will exacerbate the flood problems faced by many Third World river delta populations. Even in the north of the planet, vulnerable bits of the coast will start to slide into the sea.

Some analysts say that the Antarctic and Greenland ice-sheets could melt rapidly, producing very large sea level rises. The majority of scientists believe that it will happen much more slowly.

Impacts on rainfall patterns and likely changes in local habitats, flora and fauna are difficult to project, although a lot of research effort is attempting to do this. Many people have suggested that rainfall is likely to shift away from the equator, making, for example, southern Europe and the US Mid-West generally drier.

The possible social, political and economic repercussions of these changes should not be underestimated. The prospect of changing rainfall patterns could destabilise existing societies as pressures to migrate are increased.

Supporters of the global warming theory (sometimes described as the 'enhanced greenhouse effect') point to historical evidence to support their theory.

The Evidence

First there are the observed temperature changes over the last century or so, which can be seen in Figure 2.10.

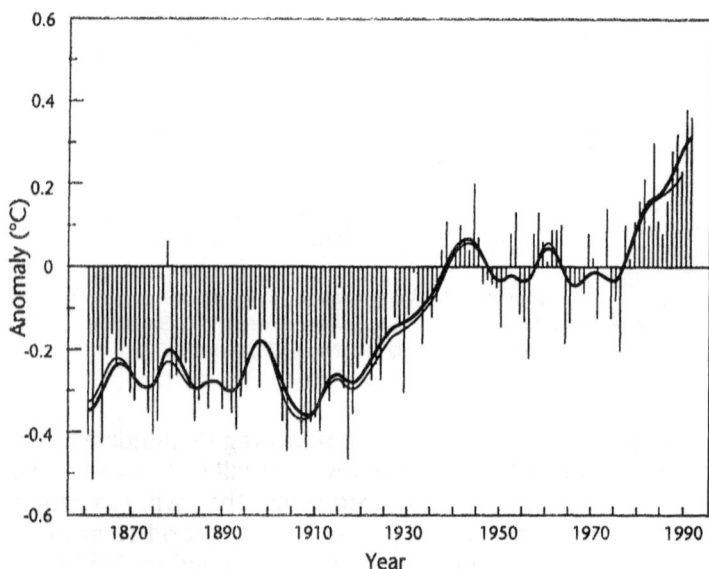

Figure 2.10 Global Changes in Land, Sea and Air Surface
Temperatures, 1861 to 1991

Source: UNEP/WMO IPCC, Callendar, Houghton and Varney (eds.), *Climate Change 1992, Supplementary Report to the IPCC Scientific Assessment* (Cambridge: CUP, 1992), p. 147

The record is of a rather uneven rise of around 0.5 of a degree over the last century. Although this does not 'prove' the global warming theory, it certainly does not contradict it.

Other historical information comes from analysis of geological evidence. Scientists have calculated atmospheric carbon dioxide levels from bubbles trapped in ice layers of known ages and compared these levels with the corresponding climatic conditions such as the disposition of ice-sheets.

Analysts like Nick Shackleton of the University of Cambridge believe that it is very difficult to model climate changes over the last 100,000 years without making the assumption that carbon dioxide levels influence the climate. His own research into the past climatic record suggests that changes in carbon dioxide levels have contributed to the growth and decay of continental ice-sheets.[12]

Criticisms of the Global Warming Hypothesis

On the other hand, there are many who take issue with the global warming hypothesis.

Analysts such as Friis-Christensen and Lassen[13] say that changes in temperature in the northern hemisphere over the last 130 years show a close correlation with variations in the length of the solar cycle. The solar cycle is the time period between peaks of sunspot activity, and shorter cycles are held to indicate increasing solar irradiance. Friis-Christensen and Lassen's estimates of the strength of the solar cycle effect seem to allow little room for any role for the enhanced greenhouse effect in influencing recent temperature variations.

Supporters of the global warming theory tend to reply that while solar cycle activity may be an influence on climate, there is no clear explanation of how the changing solar cycles affect the temperature of the Earth's atmosphere and there is no clear method of mapping variations in solar cycle lengths. Information gleaned from satellite measurements suggests that the impact of sunspots on solar irradiance is relatively small.

Nevertheless, 'greenhouse' theorists such as Kelly and Wigley have conceded that if the solar cycle effect is as strong as its supporters claim then it could explain between around 13 and 21 per cent of the temperature variance between 1750 and the 1980s.[14]

Other 'sceptics' doubt the accuracy of the IPCC's projections of climatic change. The sceptics argue that many modellers have not taken sufficient account of so-called negative feedback mechanisms which act to reduce warming. For example, some argue that greater warming will produce more evaporation and more clouds. The clouds may shut off sunlight producing a counterveiling cooling influence.

Dick Lindzen says that the real extent of warming is likely to be in the zero to 1.5 degrees range.[15]

Some people point to possible benefits of increasing carbon dioxide levels, including the faster growth rate of many types of plants.

Although the IPCC have achieved a broad measure of consensus among themselves, there is a healthy subculture of scepticism among scientists who say that the prospect and likely impact of climate change have, at the least, been exaggerated.

Those who contend that combating warming should be made a priority say that it is foolish to gamble on the impacts being smaller than currently predicted.

If the IPCC projections of warming are accurate then climatic changes will gradually accelerate during the next century unless there is a dramatic shift to stabilise global emissions within a couple of decades and then to reduce them sharply thereafter.

Some aspects of the controversy among scientists resemble the debate, in the 1970s, about the impact of CFCs on stratospheric ozone levels. Those who called for CFC use to be curbed were later vindicated by data showing ozone depletion. Could history repeat itself in the case of the impact of greenhouse gases on global warming?

Should We Take Action?

Many people say that taking action to curb the problem will be very expensive and that to take measures that would be economically crippling would be very unwise given the controversial nature of the global warming theory. Others argue that the stakes are too high to be complacent and that the acceptance of higher energy costs is preferable to the potential penalties of inaction. Still others say that many of the measures that need to be taken are low-cost measures that are justifiable on grounds other than the alleged need to combat global warming.

There is a big political problem with tackling global warming. Those who cause the problem do not suffer any direct short-term consequences. This sets the problem apart from acid rain; even when sulphur or nitrogen acids are blown across international boundaries, the arms of the polluters may be jolted by neighbouring countries which are suffering.

Thus there are barriers to implementing 'expensive' measures. What most people seem to take for granted is that tackling the issue effectively will cost a lot of money. I shall examine the truth of this assumption later.

As far as the IPCC is concerned, action certainly needs to be taken. They say that reductions of 60 per cent of present levels of carbon dioxide emissions will be needed before the climate will be stabilised. Because of the industrialised world's higher levels of carbon dioxide emissions, this means an 80 per cent reduction in carbon dioxide emissions must come from the developed world.

In order to assess current proposals for actions we need to look at the sources and trends of carbon dioxide emissions.

Trends in Carbon Dioxide Emissions

Recent trends in carbon dioxide emissions are shown in Figure 2.11. The regional sources of these emissions are shown in Table 2.5.

Table 2.5 Carbon Dioxide Production by Region, 1990 (Millions of Tonnes)

North America	5,470
OECD Europe	3,580
Africa	699
Asia + Middle East	2,491
China	2,415
Former Soviet Union	3,604
Latin America	1,025
World	22,300

Note: the region sub-totals do not add up to the world total because some countries are not covered in the sub-totals

As can be seen from Figure 2.11 global carbon dioxide emissions are rising rapidly in the world as a whole, but since the 1973 oil crisis the increase has come from non-OECD rather than OECD countries. The US is the biggest single energy-consuming country, and produces over a fifth of all man-made carbon dioxide emissions although it has only 5 per cent of the world's population. The

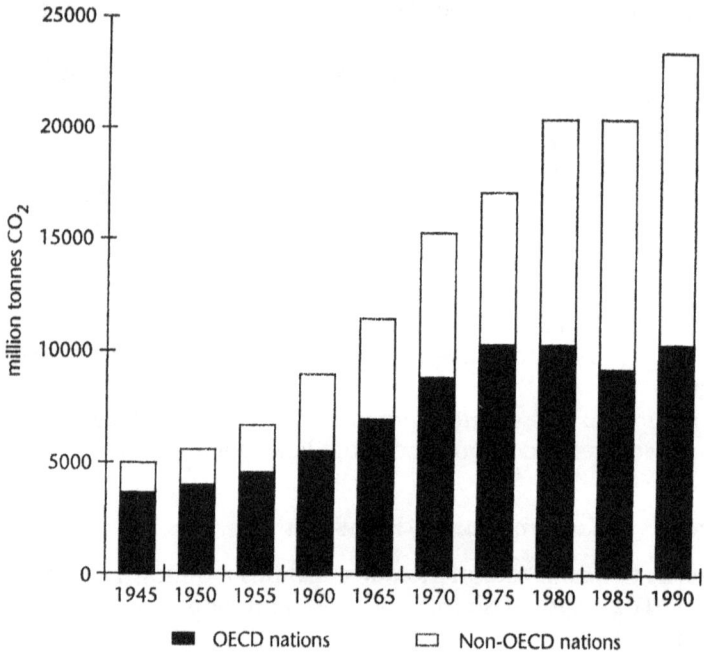

Figure 2.11 World Carbon Dioxide Emissions
(Millions of Tonnes from Fossil Fuels), 1945–90

Source: Adapted from OECD/IEA, *Energy Policies of IEA Countries,*
1991 Review (Paris, 1992), p. 56

UK produces around 2.6 per cent of global carbon dioxide emissions although it contains only 1 per cent of the world's population.

Carbon Dioxide Abatement Commitments

Although, as can be seen in Table 2.6, some commitments have been made to constrain or reduce emissions, many countries, including the UK, US and Germany, say they are relying on voluntary, market-led methods to meet the targets. Many environmentalists doubt whether the targets and the methods of reaching them are sufficient.

In order to even stabilise global emissions the leading industrialised states would have to implement major cuts in carbon dioxide emissions to cancel out the upward pressure on carbon dioxide emissions from non-OECD nations. Given that carbon dioxide emissions from OECD countries (as pictured in Figure 2.11) have been fairly stable over the last 20 years, the targets set by some leading countries of 'stabilising' emissions could be viewed as being rather unadventurous.

The issue of global warming is a central issue for environmentalists.

Table 2.6 National Carbon Dioxide Emission Reduction Targets

	CO_2 reduction target
Belgium	–5%, 1990–2000
Denmark	–20%, 1988–2005
France	stabilise at +16% of 1990 level
West Germany	–5%, 1990–2000
Greece	+25%, 1990–2000
Japan	stabilise at 1990 level by 2000
Italy	stabilise at 1990 level by 2000
Netherlands	–3 to –5%, 1990–2000
Spain	+25%, 1990–2000
Sweden	stabilise at 1990 level by 2000
UK	stabilise at 1990 level by 2000
US	stabilise at 1990 level by 2000

Source: *ENDS Report 217*, February 1993, p. 36, and other sources

Oil Spills

Roughly half of all oil produced in the world is shipped, mainly as crude oil, but sometimes as refined oil products. The problem with major oil spills at sea began in the 1960s when supertankers, carrying 250,000 tonnes (about 66 million gallons), proliferated. The *Torrey Canyon* disaster in 1967 off the coast of Cornwall in the UK was the first major spill to catch the public imagination.

The oil escapes from a rig off the Californian coast near Santa Barbara in 1969 caused a great deal of damage to the image of the US oil industry. Such was the public outcry that oil drilling off the Californian coast was banned.

The biggest oil tanker spill ever (outside the 1991 Gulf War) was the 1978 *Amoco Cadiz* accident off the Brittany coast when 230,000 tonnes went awash.

The *Exxon Valdez* accident in Cook Inlet, Prince William Sound, Alaska, lost over 40,000 tonnes of oil in March 1989. This has resulted in a $2 billion clean-up operation and several years of litigation with vast sums being claimed in compensation. The episode shattered Exxon's image.[16]

The Arabian Gulf has been drenched with oil, particularly in Iran–Iraq war of the 1980s and also during the 1991 Gulf War. The tanker *Braer* went aground on the Shetlands in 1993 spilling around twice as much oil as was lost by the *Exxon Valdez*.

Oil spills kill many seabirds, especially diving birds such as the auk. Marine mammals are sometimes also at risk. Salt marshes, coral reefs and mangroves can all be destroyed. Beaches can be affected for up to ten years. Fishermen said that salmon catches plummeted in Prince William Sound near the *Exxon Valdez* spill, even though at first they seem undamaged.

In the immediate aftermath of these tragedies there are immediate calls for something to be done to save the animals, plants and threatened beaches.

However, all is not what it seems on the television screens. True, these headline oil spills have done major damage, but they form only a relatively small proportion of total spillage. The Worldwide Fund for Nature (WWF) estimates that roughly the amount spilled by the *Braer* is routinely discharged into the North Sea every year. 'Routine' discharges come from illegal tanker discharges and spills from oil and gas rigs and refineries.[17] Broadly similar results have been given in US studies.[18]

Not only are major oil spills only a small part of the problem, but efforts to clean up have at best a marginal and at worst a negative impact. The laying of booms collects only a few per cent of the oil. Cleaning up oil-splattered birds is only occasionally likely to save them. 'When I find an oiled bird on the beach, I knock it on the head,' said one worker from the Royal Society for the Protection of Birds in the wake of the *Braer* disaster.[19]

Major controversy rages over the impact of dispersants. The first generation of dispersants, used at the *Torrey Canyon* disaster, succeeded mainly in poisoning beaches with highly toxic aromatic hydrocarbons. Beaches 'protected' by dispersants took longer to recover than the 'unprotected' ones.[20] 'Safer' dispersants have serious side-effects. Even their advocates recommend they should be kept away from beaches. Their detractors say that while surface mammals and birds may be protected, this will be at the cost of poisoning fish lower down in the water column.[21]

Environmentalists want international agreements on shipping oil to be tightened. In 1993 amendments to the international MARPOL convention (which is aimed at preventing pollution from ships) were made. These changes insist on double hulls for oil tankers, put stricter controls on routine discharges from ships and also place restrictions on dumping garbage in Antarctica. These changes will stop some accidents, although a lot of routine discharges are likely to continue.

There is also a great deal of pollution caused by land-based oil drilling operations. These operations are closely regulated in the West, but the oil industry seems to be able to do almost what it likes in other areas, especially in developing countries where the indigenous population has little political clout.

In Nigeria, for example, protests against the effects of oil operations in the densely populated Niger delta region have resulted in confrontation with the Nigerian army and many deaths over recent years. The army have defended oil companies like Willbros, a US outfit, who were laying oil pipelines in 1993, but who were forced to withdraw. The local Ogoni tribe complained about the land and water being polluted by drilling, refining and petrochemical operations and about their lands being disturbed by pipelines. The Ogoni themselves do not even have access to piped water supplies.[22]

In Russia, the widespread destruction of habitats in western Siberia has been perpetrated by the Soviet oil industry itself. These habitats are easily disturbed since the cold climate slows ecological recovery processes. Apart from the impact of the pipelines, roads and chemical discharges, an estimated 10 per cent of all Soviet oil pumped from the depths has leaked, causing widespread contamination. The new Russian industry seems no more efficient than the old Soviet industry.[23]

Electromagnetic Fields

Electromagnetic fields (EMF) exist in proximity to high-voltage power equipment and also close to office and household equipment such as computer monitors, microwave ovens and cellular phones. EMF can produce electric fields in the human body which can, on some accounts, excite ions in body tissue.[24] Some studies suggest there might be cases where these fields have caused leukaemia or brain cancer.

Fields can be experienced either by being underneath high-voltage power cables or even by being very close to computer monitors. People living next to electricity sub-stations have been especially worried.

Interest in the health effects of EMF began in the late 1960s, and in 1972 it was suspected that Soviet workers near high-voltage switching stations were suffering from a variety of symptoms. More recently, research conducted in Denver, Colorado, suggested that there was a relationship between childhood cancers and people living in houses close to power distribution lines. Other studies in the US and Scandinavia have suggested similar links.

At first these conclusions were strongly contested by other scientists, with the UK's National Radiological Protection Board (NRPB) discounting any carcinogenic hazard except where people were doing welding work or working next to television transmitters or electric furnaces.[25] However, soon after, the chairman of the study group that produced the NRPB report, Sir Richard Doll, conceded that there could be 'a small risk that was spread very widely'.[26]

The Swedes have proposed that schools and other facilities for children should not be sited in areas with high magnetic fields and have set regulations about EMF levels. Companies such as Apple are bringing their computers into line with these regulations. In the US, electricity companies such as Southern California Edison are installing EMF protection equipment. Around 3 to 4 per cent of the capital costs of the company's electricity sub-stations are related to anti-EMF measures.[27] New power lines are being routed to avoid houses.

Many US lawyers are already anticipating large compensation claims for alleged EMF-related sickness. EMF worries may also be added to the long list of complaints about new power developments.

EMF is an issue that still has a long way to run.

3

The Resource Problem

Problems occur when a resource becomes in such short supply in relation to demand that its price rapidly and markedly increases.

Perhaps the first major national energy resource crisis occurred in the UK in the sixteenth and seventeenth centuries. The shortage of trees was such that there was not enough timber to sustain domestic armaments production. This crisis was resolved as coal took over the domestic and then the industrial energy sectors. As the industrial revolution spread from Britain to the rest of Europe, so did coal use. World coal consumption increased from 15 million to 700 million tons during the course of the nineteenth century.[1]

The first 'rock' oil was extracted in Pennsylvania in 1859 to feed the oil lamp industry which was suffering from a shortage of whale oil.

The internal combustion engine was fully developed by the end of the nineteenth century and oil consumption grew rapidly. Petroleum and the motor vehicle made an ideal marriage. Petroleum is a very compact, dense energy source and the internal combustion engine is built at a small size without the need of the boiler associated with solid-fuel-fired steam engines.

During the earliest stages of the oil industry, oil prices were highly volatile, 'crashing' and 'spiking' in quick succession. Then Rockefeller gained dominance over the industry and stabilised prices through his much-maligned monopoly practices. Although his Standard Oil company was eventually broken up by federal order, the heirs to the Rockefeller empire controlled and planned the development of supply so that the supply of oil was matched to demand. Price stability was achieved.

Whereas the UK blazed the industrial trail with coal in the eighteenth and nineteenth centuries, the Americans achieved world dominance with oil in the twentieth century.

After the Second World War oil advanced in tandem with the growth in motor vehicle use. The increasing demand for gasoline meant that other fractions of the oil drawn off during the refining

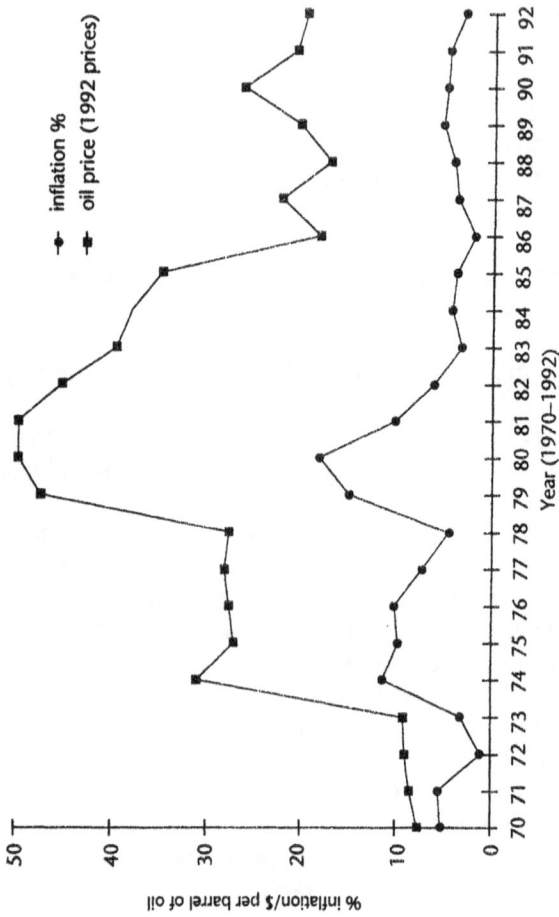

Figure 3.1 Oil Prices and US Inflation, 1970–92

Source: California Public Utilities Commission, *California's Electric Services Industry* (San Francisco, 1993), and *BP Statistical Review of World Energy* (London, 1993)

process could be sold cheaply to commerce and the home as fuel oil and heating oil.

Oil displaced coal as the world's main fuel, and during the 1960s oil demand was hurtling upwards at an incredible rate. Oil dominated, or was threatening to dominate, most energy markets, from heating oil to power stations, from transport to industrial process heat. The oil industry was the largest industry in the world, and as long as price stability reigned, its continued advance seemed assured.

However, price stability disappeared in the 1970s. Oil-producing countries were becoming more independent and assertive. OPEC, the Organisation of Petroleum Exporting Countries, was formed in 1960. Many countries nationalised their country's oil assets, thus reducing the power of the oil companies. By the early 1970s, the US was a big net oil importer and was unable to increase production levels further. The oil companies were forced to bargain seriously with OPEC who, by the early 1970s, were producing half the world's oil.

Oil prices began to go up before the 1973 crisis, but after the outbreak of the Arab–Israeli Yom Kippur war in November 1973 the Arab states mounted an oil boycott of Western states that supported Israel. This triggered a sudden fourfold price rise and made the oil companies' attempts to bargain over oil prices irrelevant. Inflation rocketed throughout the West, as can be seen in Figure 3.1.

Increases in oil prices lead to increases in other energy prices. Gas prices, for example, are often linked directly to oil prices. The price rises cause inflation, and governments tend to put up interest rates in an effort to choke off the inflation. High interest rates can lead to recession.

Strikes of oil workers in revolutionary Iran sparked off a second oil price hike in 1979. Oil prices jerked upwards yet again after the start of the Iran–Iraq war in 1980. During the early 1980s OPEC managed production levels in order to keep up oil prices. However, their control loosened in 1985 as some OPEC members consistently produced more oil than they were allowed under OPEC quotas. Oil prices declined sharply in 1985 and, apart from a small 'spike' in prices at the time of the Gulf War with Iraq, oil prices were low in the late 1980s and the first part of the 1990s.

During the 1970s many believed that the oil crisis was symptomatic of a far wider depletion crisis. Nevertheless, the notion of imminent global depletion of oil supplies proved to be a myth. Academics like Peter Odell challenged the gloomy 'energy shortage' prognosis. Although two-thirds of the world's oil reserves are in the Middle East, Odell and others claimed there was much oil to be discovered outside the Middle East. Official oil reserves amounted to 30 years of global consumption in 1973. In 1993 reserves stood at 43 years.

Table 3.1 Changes in Global Energy Demand, 1979–92 (Million Tonnes of Oil Equivalent (MTOE))

	1979	1985	1992
Oil	3,225	2,809	3,146
Natural gas	1,344	1,479	1,759
Coal	1,969	2,083	2,153
Nuclear power	155	369	541
Hydroelectricity	145	170	190
Total	6,838	6,910	7,789

Note: the figures for nuclear power are artificially inflated compared with hydro-electricity. The figures for consumption of nuclear power include all the heat produced by nuclear power stations of which only a third is turned into electricity

Source: *BP Statistical Review of World Energy* (London: BP, various years)

Oil demand has stagnated and oil-consuming nations have discovered alternatives to politically sensitive Middle East oil. There has been an expansion in non-OPEC oil production. There has been an increased emphasis on energy efficiency. Industrial, commercial and domestic consumers have switched away from oil and, increasingly, towards natural gas. (See Table 3.1 for changes in energy demand since 1979.)

These days analysts generally ascribe the causes of the oil crises to political factors. This is true, but only half of the picture. The other half is the increasing tendency towards energy dependence. There is no imminent shortage of fossil fuels (coal, oil and natural gas) on a global level, but there is a mismatch between energy suppliers and energy consumers. The major energy consumers tend not to be able to produce enough oil and, increasingly, natural gas to match their own rising consumption. The classic example of this is US oil production which peaked in 1970 because of the exhaustion of many of the cheaper oilfields. The level of US oil production seems likely to continue its decline (see Figure 3.2).

Serious exploitation of UK oil reserves began only in the 1970s and, to date, production has continued to increase in fits and starts.

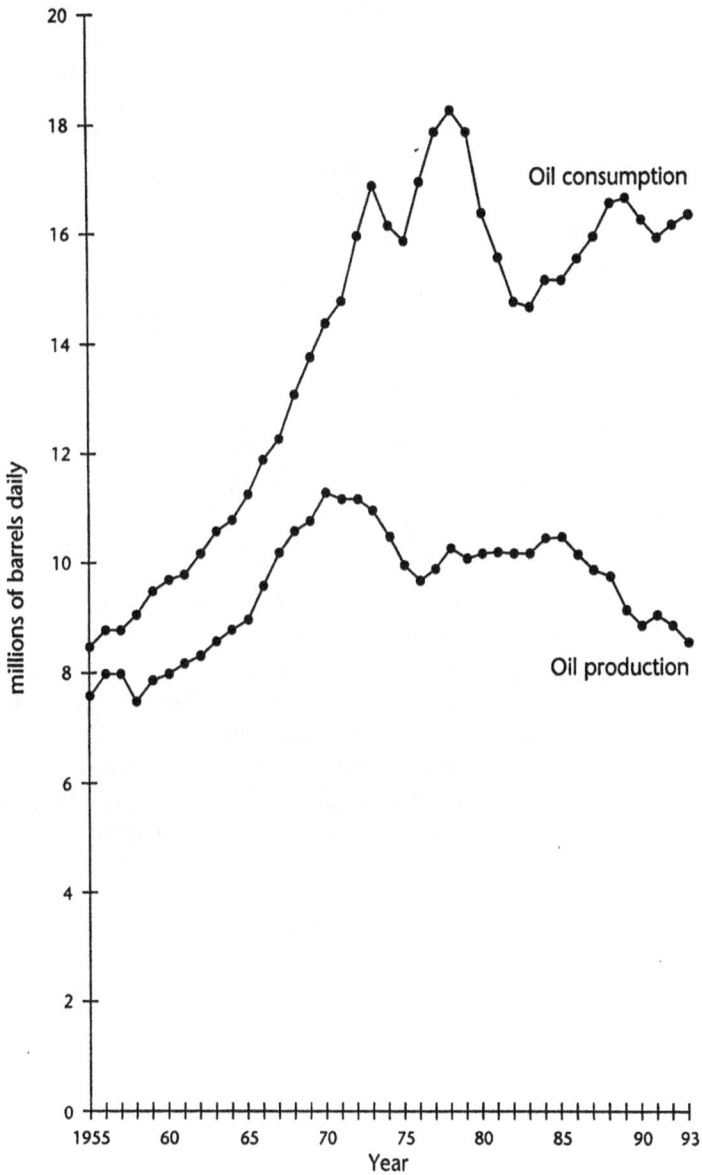

Figure 3.2 US Oil Production and Consumption, 1955–93

Source: *BP Statistical Review of World Energy* (London, various years)

This will not continue indefinitely. It is not a question of whether the UK's oil production will decline, but when. So will there be energy resource crises in the near or medium-term future?

In the past few years declining Russian oil production has meant that once again the West is importing larger quantities of Middle Eastern oil, as can be seen in Table 3.2.

Table 3.2 World Oil Production (MTOE)

	1979	1988	1992
OPEC	1,553	1,037	1,284
Former Soviet Union	586	624	450
Rest of world	1,086	1,387	1,436
Total	3,225	3,048	3,170
OPEC as % of total	48	34	40.5

Source: *BP Statistical Review of World Energy* (London: BP, various years)

There are two opposing analyses of the future of oil markets. The first is the 'capacity crisis' thesis. This says that investment in new oilfields and refineries is trailing off so that there may not be sufficient capacity to meet increasing demands for oil coming from the rapidly expanding Pacific Rim and a revived West. Some commentators have said that this could lead to a new oil crisis.[2]

Oil companies (supported by many oil analysts) say that the high extra cost of meeting new environmental regulations to clean up oil extraction, refining, shipping and distribution activities increases the prices of petroleum products and deters investment in new oil supply capacity. Others, including Peter Odell, say that continued low demand for oil and the continued availability of non-OPEC oil will stave off any capacity crisis, unless it is sparked by a political crisis.[3]

Greenpeace contests the claim that environmental improvements will lead to major oil product price increases. For example, Greenpeace claims that confidential estimates prepared for the Shell oil company say that meeting new international regulations for oil tankers will cost only 3 to 5 cents a barrel. This would have a negligible effect on fuel prices.[4]

Some analysts fear that, in the absence of any force that can stabilise oil markets, price volatility will result. Oil price 'spikes' may be sparked by political crises in key oil-producing states. The more energy-dependent a state is, the more it is vulnerable to sudden increases in the price of oil.

OPEC states have become used to the large incomes generated by oil exports. This income has declined in recent years and political instability may result as the economies of oil-producing states suffer. The worst-case scenario might be instability in Saudi Arabia which produces an eighth of the world's oil.

An important reason why oil prices have sagged since the mid-1980s is because of the increasing share of world energy demand supplied by natural gas. Increasing reliance on this fuel is one means of reducing dependence on oil imports. However, as natural gas demand expands many nations are becoming dependent on imports. Oil dependence is being replaced by gas dependence. Could the future involve a gas resource crisis? This issue will be examined further in Chapter 6.

The point that should be emphasised is that action to reduce reliance on fossil fuels can both reduce pollution problems and reduce the energy dependence that can bring with it a host of economic and political hazards. In other words, environmentally sensitive energy policies can act to mitigate conventional problems as well as ecological concerns.

4

Solutions

You have read about the problems. The rest of this book is concerned with solutions.

Solutions can be categorised under four general headings.

The first, and perhaps most familiar type of solution, is a so-called *'end of pipe'* measure involving fitting equipment that will remove a particular pollutant. For example, coal-fired power stations can be fitted with equipment to remove sulphur and nitrogen emissions, and motor vehicles can be fitted with catalytic converters to remove various pollutants.

However, 'end of pipe' solutions, while essential in many circumstances, are often limited in their scope. They are rarely 100 per cent effective, and thus their effect is diluted by increasing energy use. There are two further problems with 'end of pipe' methods. One is that they do nothing to ameliorate fossil fuel resource problems. Another is that they usually involve increases in costs, even if these costs are sometimes overstated by the energy suppliers.

The second type of solution is *fuel switching*. This involves substituting one fuel for another. It can have the advantage of reducing pollution and in many cases it can act to reduce resource problems. For example, nuclear power, renewable energy and natural gas have been substituted for coal and oil in power stations since the 1970s. This has reduced SO_2 and NOx emissions. Switching from coal and oil to nuclear power, renewable energy sources and, to a lesser extent, natural gas, can also reduce carbon dioxide emissions.

However, fuel switching can also create its own problems. Different types of pollution can be produced. Nuclear power, for example, produces nuclear waste. Renewable energy sources, which include hydroelectricity, wind power, biofuels, solar power, tidal power, geothermal energy, wave power and other technologies, have their own environmental consequences.

Following the 1970s, coal was substituted for oil which had dramatically increased in price. This enabled many countries to reduce their oil resource problems. However, some states, in particular East Germany, used unclean technology with which to burn the coal, with disastrous environmental results.

A third type of solution is that of improving '*supply-side efficiency*'. This involves reducing the waste involved in supplying energy to the final consumer. The biggest energy wastage occurs when electricity is generated. Between 50 and 70 per cent of the energy used by currently operational power stations is not turned into electricity. Consequently the amount of pollution produced in the course of electricity production is at least twice and maybe three times the pollution that would be produced if most of the energy input was turned into electricity. Much or most of the otherwise wasted energy can be used to provide heating services in combined heat and power or cogeneration systems.

A fourth type of solution is that of improving the efficiency with which a given amount of delivered energy is put to use. Thus is called '*end use efficiency*'. Buildings can be designed to use only small amounts of energy to provide good levels of heating. Electrical equipment such as refrigerators can be designed to use less energy to produce the required amount of cooling. Much of the energy used by the computer wordprocessor on which I am working goes on heating up the circuits rather than helping me write this book!

Energy efficiency involves no pollution or resource problems, but although interest in energy efficiency has increased over recent years there is widespread doubt about the cost and practicality of implementing it.

There is tremendous debate about the cost and desirability of these different strategies. The main divisions in this debate emerged in the 1970s. Until then there was widespread agreement that in the future our energy needs would be increasingly met by nuclear power. The oil crises of the 1970s should, in theory, have entrenched this view, but instead it was increasingly questioned by ecologically oriented analysts such as Amory Lovins.

Lovins characterised the conventional view as the 'hard' path involving centralisation and the dangers of radioactivity. Lovins, who acted as an adviser to US Friends of the Earth, talked about an alternative 'soft' path involving energy efficiency and renewable energy.[1]

Increasing attention was given to environmental problems. The first programmes aimed at countering acid rain and reducing pollution from motor vehicles were put in place.

In the mid-1980s the terms of the debate began to change again. Oil prices came down and the danger of fossil fuel depletion seemed to recede. Then, on the pollution front, concern about the impact of carbon dioxide emissions on the Earth's climate grew. Paradoxically, this new ecological concern encourages the idea that nuclear power may be necessary for environmental reasons. The use of nuclear power can substitute for fossil fuels, so reducing carbon dioxide emissions.

During the 1980s the concept of *'sustainable development'* was widely elaborated, in particular by the Brundtland Report, published in 1987. Sustainable development involves improving living standards using means that can be continued indefinitely without compromising the ability of future generations to sustain improvements in their own living standards.

In many ways the divisions drawn up during the energy debates of the 1970s remain the same today. So-called 'greens' want to reduce pollution caused by fossil fuels through improving the efficiency with which energy is used. They also tend to favour, with varying degrees of enthusiasm, the use of renewable energy. Other, more conventional approaches, stress that nuclear power will be needed if energy supplies are to be guaranteed and air pollution reduced.

Behind all the arguments about what constitutes the most appropriate shape of a sustainable energy strategy is the issue of costs. Consumers want solutions to environmental problems, but they also want solutions that are going to cost the least. Thus it is vitally important to cost the various solutions using reliable cost criteria. Only then is it possible fully to evaluate the worth of the various solutions. The next chapter concerns costing energy production and use.

5

Costing Solutions

Environmentalists want to save the planet from pollution caused by energy sources. They have persuaded the public that it needs to be saved. Yet the public want it to be done as cheaply as possible.

The ideal solution would be to save the planet and save money at the same time. In some instances, such as retrofitting existing fossil fuel equipment with scrubbing equipment, this is unlikely to be the case, but in other instances, such as improvements in energy efficiency, it might happen. The point about energy efficiency is that it gives us the possibility of reducing the quantity of energy needed to provide a given level of service. Energy efficiency can certainly cut pollution, but can it really be achieved without increasing energy bills, and can it cut pollution by sufficiently large amounts to make a real impact on resource and pollution problems?

What about the costs of non-fossil (nuclear and renewable) energy sources? At the present time these seem to need subsidies. Are there any signs that such alternative sources could in the future compete on commercial terms with fossil fuels without special help?

We shall be in a position to answer these questions only if we can find a common, reliable way of costing the various energy options.

The Environmental Economics Approach

A central problem with ecology is that it is concerned with impacts on the environment that are usually not quantified in monetary terms.

Environmental economists have attempted to cost the effects of pollution and different ways of cutting pollution. The techniques involve the conversion of the costs of environmental damage into monetary values. These costs are then added on to the conventionally assessed costs of supplying energy to produce a 'real' cost, to society, of energy consumption.[1]

Few people deny that our energy activities cause damage to society, flora and fauna. Whether it be damage to crops by ozone, or respiratory disease caused by dust or sulphur particles, loss is caused that reduces our wealth. However, the energy consumer does not pay for these costs. Because they are not included in the commercial energy transaction, such costs are called external costs.

Broadly speaking, there are three means of attaching monetary value to environmental damage. Some try to calculate the cost of the equipment that needs to be fitted in order to remove pollutants (for example, flue-gas desulphurising equipment). Others work out the costs of changes needed so that we can live with the pollution (for example, irrigation to compensate for lower rainfall after global warming). Still more calculate the economic costs of the damage caused by the pollution (for example, damage to crops from ozone or acid deposition).

The trouble is that these different methods tend to produce widely different results. Moreover, the external costs of some environmental impacts, for example the loss of countryside caused by open-cast coalmining, the disappearance of butterflies because of acid rain or the loss of wetlands resulting from sea level rises caused by global warming, can be imputed only by indirect means. People have to be asked in surveys how much such things are worth, or they have to be asked what they would pay to avoid such losses. The scope for disagreement about these costs is very wide indeed.

More arguments ensue when we try to assess those external costs of our energy consumption that are borne by other countries. Sea level rises caused by global warming could worsen the impact of flooding in Bangladesh, for example. The average person in the industrialised world is responsible for pumping out a lot more carbon dioxide into the atmosphere than the average Bangladeshi. Yet because the income of the average Bangladeshi is much lower than the per capita income of people in OECD countries, the economic loss suffered by the Bangladeshi will tend to be given a low value. This is an extremely perverse result. There are broadly similar problems when one tries to assess the external costs of damage to other species and to future generations.[2]

On top of this there are disputes about the extent and impact of phenomena such as global warming. It is hardly surprising, therefore, that monetary valuation techniques produce different results. Andrew Stirling has prepared a chart (see Figure 5.1) showing these vast differences in the case of pollution coming from various electricity sources. The differences vary by a factor of 50,000.

Decimal log scale
¢/kWh (1988)

| 0.001 | 0.01 | 0.1 | 1 | 10 | 100 | 1000 2000 |

Biomass

Onshore wind

PV

Nuclear fission

Gas

Oil

Coal

| 0.001 | 0.01 | 0.1 | 1 | 10 | 100 | 1000 2000 |

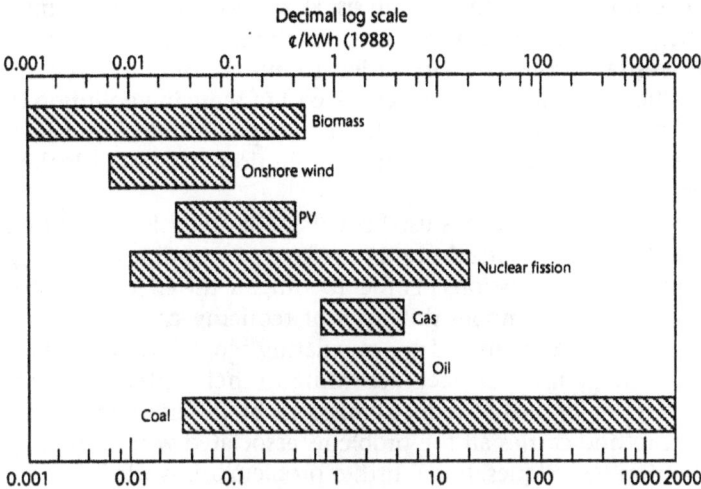

Figure 5.1 Overlap between Ranges of Environmentally Related
Damage Costs for Selected Electricity Supply Technologies

Source: Andrew Stirling, 'Regulating the Electricity Supply Industry by
Valuing Environmental Effects', *Futures*, December 1992, pp. 1024–47

Stirling attacks the usefulness of monetary valuation techniques. He concludes that rather than acknowledging the limitations of economics in dealing with environmental issues economists have sought to colonise the area, producing, in the process, inaccurate results posing as precise scientific judgements. He criticises the implication that decisions ought to be delegated to technocratic elites whose deliberations are opaque to public scrutiny. Stirling suggests that environmental priorities should be determined by public debate rather than by numerical calculations made by environmental economists.[3]

Monetary valuation techniques do have important uses. They can be valuable as a way of demonstrating that energy and other human activities can have significant impacts on the quality of life that are not quantified in commercial transactions. The trouble is that sometimes monetary valuation techniques are used to recommend lines of political action. This is especially the case when monetary valuation techniques are used as part of something called cost/benefit analysis.

The aim of cost/benefit analysis, as applied to environmental issues, is to calculate what amounts should be spent on environmental protection in order to maximise our welfare. Cost/benefit analysis compares the cost of reducing pollution to the benefits to society that result from the pollution being abated. An optimum level of spending on pollution abatement is then calculated.

William Nordhaus has used cost/benefit analysis, in the case of the US, to determine the optimum level of spending on abating carbon dioxide emissions in order to mitigate the effects of global warming. In his model the costs of reducing carbon dioxide emissions are determined by calculating what level of carbon taxes (energy taxes on fuels according to their carbon content) needs to be applied to reduce emissions by different amounts.[4] This method carries all the problems associated with monetary valuation techniques, but a further problem arises with the way Nordhaus uses energy taxes as a means of estimating the costs of reducing emissions.

A different approach might argue that many measures, especially energy efficiency measures, actually reduce the consumer's cost of enjoying a given level of energy services. Energy taxes increase consumer costs and are not necessarily the only means of reducing energy consumption. If cost/benefit analysis was used properly then it would recognise that some reductions in emissions could be achieved at negative cost.

The efforts of environmental economists to determine the costs of pollution from energy sources and assess the costs of combating this pollution seem to be flawed in many respects. Let us now look at a different approach that avoids these pitfalls.

The Cost of Saving the Planet

The reduction of pollution through 'end of pipe' measures usually involves extra costs, although even here costs are often smaller than anticipated.[5] However, fuel switching and, especially, energy efficiency measures can often lower energy costs and can be an effective way of reducing many types of pollution.

In the case of reducing carbon dioxide emissions the very nature of the problem puts a premium on reducing the amount of fossil fuels we use. This presents us with an opportunity to reduce costs by using fossil fuel energy more efficiently. For example, my interest in environmental matters made me aware of the large energy losses going up my fireplace chimney. I blocked it off with a piece of plywood (while leaving a small hole to avoid condensation problems). This stopped the loss of heat up the chimney and significantly reduced my energy loss and my heating bill. My investment in the wood paid off extremely quickly. I save money and emissions of various sorts are reduced.

Some might regard this as a trivial example, but it is one that illustrates how a shift in cultural attitudes towards environmentalist aims can promote the take-up of cost-saving measures that might otherwise have been missed. Large companies such as IBM now regard the achievement of energy-saving objectives as an important part of their corporate image.

The problem of cutting pollution from energy sources becomes one of changing institutional arrangements to ensure that society invests in conserving energy resources as well as in supplying them.

Many conventional economists dismiss the idea that energy efficiency can reduce energy consumption unless energy prices are increased to make people keener to take up energy-saving measures. Indeed, some even propound a sort of 'iron law' of energy consumption which says that money saved through energy efficiency measures will be spent on energy somewhere else.

Increased knowledge (for example, a wider appreciation of the advantages of energy efficiency) can substitute for natural resources and can allow them to be used more efficiently.[6] It does not necessarily follow that more economic development will result in a proportionate increase in the use of natural resources.

Spending less money on an energy intensive activity can be replaced by spending more money on a low energy activity. Economies can double their output, but if the goods and services are manufactured using only a third as much energy per unit of output then the economy will cut its overall energy consumption by a third.

What we want are energy services such as lighting and heating, not energy. This is a distinction that is ignored by many conventional economists who dabble in energy economics. The difference is better understood by engineers and physicists who study the use of energy. The same energy service can be delivered by combining lower energy consumption with greater investment in energy-efficient technology.

This can be achieved without any increase in costs if the cost of investments in energy efficiency is no greater than the cost of the energy use that is avoided. The energy efficiency investments can be financed by borrowing so that the loan repayments are covered by income from energy savings.

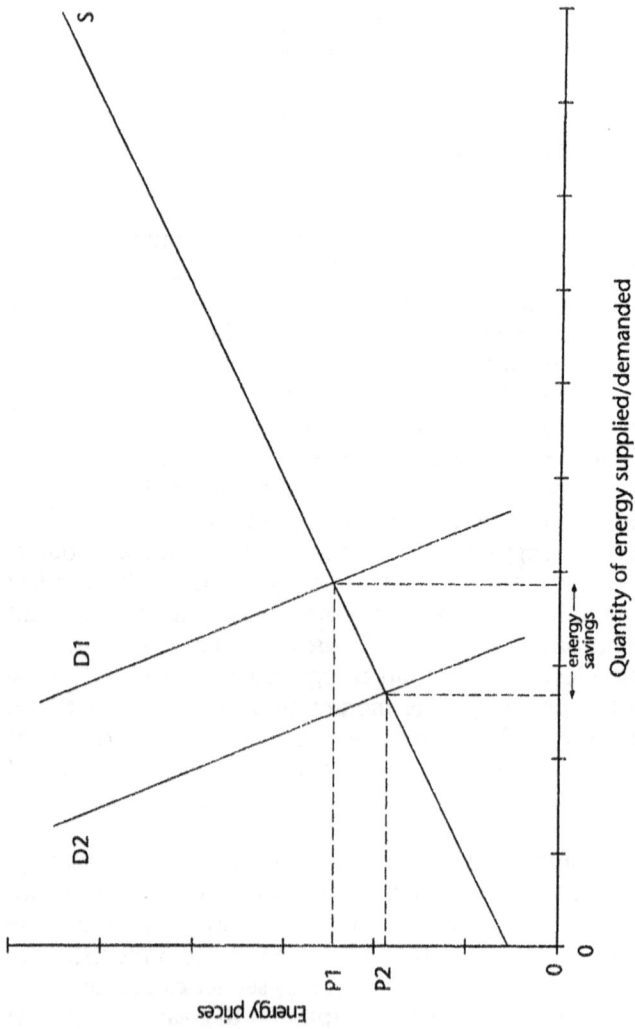

Figure 5.2 Demand for and Supply of Energy
Impact of energy efficiency measures

Because the economists who deploy monetary valuation techniques and cost/benefit analysis habitually assume that energy taxes are the main way of promoting the efficient use of energy, the impression is given that employing energy-saving techniques to reduce pollution increases energy bills. Hence the energy-saving measures come out looking much more expensive than they really are, and the optimum level of pollution reduction is often assessed to be relatively modest.

Unless energy taxes are levied at high levels they are usually an ineffective means of encouraging investment in energy efficiency because of the low elasticity of demand for energy with respect to price. This means that for a given change in the price of energy there is only a relatively small change in the demand for energy. On the other hand, this low elasticity of demand for energy will ensure that regulatory means of encouraging investment in energy-efficient technology (such as energy efficiency standards) will produce energy savings and a lowering of consumers' bills.[7]

Environmentalists often highlight the fact that consumers' energy bills can be reduced (following energy efficiency investments) because less energy needs to be bought. However, less attention has been drawn to a second financial benefit: the price that the economy as a whole has to pay for primary energy supplies will be lower.

This can be seen in a demand and supply chart (see Figure 5.2). This is an extremely simplified version of reality since the energy market is extremely fragmented, but it illustrates how a change in tastes can move the 'demand for energy' curve (D1) to the left (D2) so that for a given energy price less energy will be consumed. Such a change in tastes can come about through a change in corporate or personal attitudes or through interventionist measures to ensure that there is greater investment in energy-saving equipment.

Because the energy demand curve is shifted to the left it cuts the energy supply curve (S) at a point where less energy will be supplied at a lower price. The equilibrium energy price is reached at a lower level (P2) than would have been the case without the energy efficiency measures (P1). In other words, energy efficiency measures can act to reduce the prices of 'raw' (primary) energy supplies as well as saving energy.

The emphasis put on the need to price fossil fuel use 'properly' to account for external costs (this usually involves proposals for carbon taxes) distracts attention from the potential cost-effectiveness of energy-saving measures. If energy taxes are levied as the means of encouraging energy efficiency, the average consumer suffers an increase in energy bills. On the other hand, if the energy industry is regulated to invest in cost-effective energy efficiency measures, the average consumer should enjoy a reduction in bills.

Our judgement over whether and how highly energy ought to be taxed should be guided not by paper assessments of external costs, but on whether the taxes will achieve precisely defined policy objectives. The issue of energy taxes will be discussed further in Chapter 7 where I evaluate the different measures that can be taken to promote the efficient use of energy.

The point I want to make now is that many economists tend to confuse markets with perfectly competitive markets. They think that society's best interests will be served if the right price signals (incorporating external costs) are sent to the market. But energy markets are not perfectly competitive markets, no matter how competitive the supply side of the market may be. In order to have a competitive market there must be equal competition. Yet energy producers (whose interests lie in selling energy) and energy consumers (whose interests lie in using energy efficiently) do not have equal access to either information or capital.

The energy producers know all about energy supply while the energy consumers have little knowledge of how to use energy efficiently, and little time or money available to find out. Even more crucially, energy consumers will invariably require a much quicker payback on their investments than energy suppliers. Many individuals and companies have big debts and pressing short-term commitments. While an energy consumer may need to have his investment paid back in six months, the energy producer will have access to financial arrangements that will allow him to wait seven years or more for his original investment to be fully recovered.

In recent years many have argued that deregulation of energy supply markets, which abolishes the traditional monopolies enjoyed by energy utilities, will encourage greater competition between energy suppliers and force them to offer energy efficiency services.

There is no evidence in the sectors of the British electricity and gas markets that have so far been deregulated that this is happening to any significant extent. It is unlikely that it will happen for the simple reason that energy suppliers have to give the shareholders and banks a return on the money they lent to the energy suppliers for the purpose of building the power stations, gas pipelines and so on. The energy suppliers will try to sell as much energy as possible, not borrow more money to invest in energy efficiency!

These various market failures shift the balance of investment towards energy supply, despite the fact that the optimal mix of resources for society as a whole would involve much greater investment in energy efficiency and much less investment in energy supply. There needs to be intervention in energy markets to ensure that there is a mainstream, not just a fringe, market in energy efficiency services. This can produce greater, not lesser, competition. The tools used by environmental economists do not seem to recognise these structural problems.

I shall therefore not use monetary valuation or cost/benefit analysis to determine the costs of energy solutions; they are of questionable accuracy and lead to misleading policy conclusions. As Andrew Stirling has suggested,[8] many of these tools represent an attempt by environmental economists to impose a sort of intellectual colonialism on the issues and to reserve decision-making to an elite. Decisions must be made by informed public debate.

Instead, I shall cost the various energy options using standard commercial criteria without regard to their external costs. The costs of the options can then be compared with one another and an attempt made to determine which mix of options offers the cheapest path to the attainment of democratically determined environmental objectives. This discussion should also encompass the most cost-effective and convenient means through which these options can be implemented.

It may be that the least-cost mix of these 'environmental' options is in fact cheaper than traditional means of supplying energy. It could be that the meeting of environmental targets will actually lower, not increase, the energy bills that have to be paid by the consumer. The first step, therefore, is to determine a basis for assessing the costs of supplying energy services to the consumer using standard commercial criteria.

The Commercial Cost of Energy

I shall use private sector, commercial, standards in order to set the groundrules for assessing costs. Some environmentalists might say it is wrong to put market considerations first. Yet in the real world, when the solutions are implemented, they are going to have to be implemented in market conditions. Businessmen *have* to choose the cheapest options. In fact, public sector bodies are really in the same position. If they choose politically correct (I use the term in its general sense) but expensive solutions to energy problems they will have to cut back on educational, recreational and other services.

The methods used to cost ways of producing energy are explained below.

Costing Energy Production

The costs of producing energy are usually given in measures such as £/GJ or cents/kWh.

The cost of producing energy from a particular source is usually broken down into three parts: capital costs, which are the costs of the machinery and plant; fuel costs, which vary with output; and running costs. The latter involves mostly cleaning and repair.

Computing the fuel cost is straightforward. You merely divide the cost of fuel used in a given period (say a year) by the energy produced in that period. If a plant uses an expensive type of fuel then its fuel costs are going to be high unless it uses the fuel very efficiently. In an electricity system the plants with the highest fuel costs will generally be reserved to provide power at times of 'peak' or maximum demand. Thus 'peak' power is expensive.

Running costs are also straightforward. The costs of organising, cleaning and repairing the unit for a year are again divided by the energy produced in that year. Some types of power plant need to spend much more on cleaning and other operational activities than others. Power stations that need expensive repairs (as they all do eventually) may be closed down if new power stations can supply cheaper power.

Capital costs are trickier to deal with. While payments for running costs and fuel costs can be made as the fuel and services are required, capital usually has to be borrowed before the start of the project and before any revenues are received for the energy that the project produces.

Capital that has been borrowed is usually repaid in stages along with interest charges. These payments are usually structured so that they take place over a long period of time. If a contract exists to supply energy over the long term then the project's annual capital repayments will usually be stretched over the length of the contract. The annual repayments are then divided by the annual energy production to produce an annualised capital cost.

The annualised fuel, running and capital costs are then added together to produce total production costs.

However, there are some further complications, and these mainly involve the capital cost element. The problem is that capital has to be borrowed from somewhere.

In practice, the capital for energy projects usually comes partly from bank loans and partly from equity. Equity is cash provided by shareholders. The banks want their money back plus interest while the shareholders are paid a dividend which comes from company profits. The returns (before tax) to the shareholders have to be quite large since the shareholders not only want their original money back quickly as well as a profit, but they also have to pay tax on their dividends.

Thus capital costs consist of loan repayments, loan interest charges and returns to the shareholders. These capital costs will in theory be larger in the first years because of high interest charges on the debts that still have to be repaid, but in practice banks 'amortise' the bank loans and interest payments. This produces an arrangement where each year involves the same annual or quarterly payment to the bank(s) for as long as the loan is set to last. This is like a mortgage.

Fossil fuel energy production is often cheap because the capital cost of the equipment is relatively cheap. This is not surprising: equipment used to burn fossil fuel has had its performance improved and its costs reduced over a period of centuries. On the other hand non-fossil energy sources are usually more expensive because the capital costs (per unit of output) are much higher than those of fossil fuel plant. Non-fossil sources usually have low or zero fuel costs except for some types of biofuels. However, without high increases in fossil fuel prices this is usually not enough to compensate for the difference in capital costs.

Of course, if a non-fossil power plant has paid off its capital costs then it is often cheaper than fossil fuel plants. This is sometimes cited as a reason why nuclear and renewable energy systems should really be taken as being cheaper than fossil fuels. The trouble with this argument is that while it may be eventually cheaper people tend to want their returns in the short term, not the long term. This may be described as being shortsighted, but there is economic sense there. The point is that if you can be repaid your original investment in a shorter time then you can re-invest money in other useful (preferably environmentally sensible) things. Alternatively, the energy can be sold at a cheaper price while still satisfying the investor. The challenge for non-fossil fuels is to reduce their capital costs and increase their output.

The high capital costs of nuclear power is a crucial factor which explains why no new nuclear power plants have been ordered for a long time in the US, where energy utilities have had to finance their nuclear programmes through commercial borrowing.

The speed with which investors want their money back is governed by something called a discount rate, a widely used concept in economics and business. Indeed, the phenomenon of discounting has wide-ranging consequences for the economic organisation of society.

Discounting

We all use discounting to decide how we are going to spend our money, or, if we have enough of it, how we are going to invest our cash. Now, we do not quantify our discount rates, but we use them implicitly. For example, are you going to make a bulk purchase of, say, washing powder now, that will save money in the future or are you going to buy a small packet now, which may be more expensive in terms of washing powder but will leave you more to spend in the short term? To put it in economic terms, what is the 'opportunity cost' of your investment? What returns do you lose by not investing your money in something else?

People will prefer a given amount of money now to the same amount in the future. Often, when asked to explain this phenomenon, they will say that inflation reduces the value of money. It does, but this is only a part of the explanation. The point is that life is short and even the medium-term future may be uncertain. Moreover, if you are short of money you can ill-afford to go without in the short term in order to benefit in the future.

Thus, if you are poor or even an average middle-classer weighed down by the running costs of your affluent lifestyle, the prospect of receiving a given amount of cash in a year's time will be worth much less than being given the same amount of cash now. In other words, the value of future income or income that results from savings in costs made in the future declines the farther ahead in time that the income is to be received.

To give an example: if £100 of income gained in a year's time is worth only £75 now (excluding the effects of inflation) then you are working on a discount rate of 33 per cent (100 divided by 1.33 = 75). This devalues the present worth of future earnings very quickly. A given amount of income now will be worth less than a quarter of its present value, at a 33 per cent discount rate, if it is received in four years' time ($1.33 \times 1.33 \times 1.33 \times 1.33 = 4.21$).

Companies will assess investments by discounting the value of the future income receipts (for example, from energy sales) to see what they are worth today. This is called the present value. You can alter the discount rate you use until the present value of the future stream of income exactly equals the original investment (the net present value is said to be zero in this situation). This will give you an internal rate of return (IRR) for a particular project. If the IRR is higher than your own 'test' discount rate then the project could be a good runner.

Energy supply projects such as large power stations will usually be assessed at much lower test discount rates than are used by most energy consumers. This is because of the secure nature of the contracts, the reliability of the technology and the favourable terms upon which the project can be financed. Combined cycle gas turbine projects in the UK, for example, have been assessed at test discount rates of about 10 per cent. This is much lower than the rate used by most energy consumers to assess the viability of energy efficiency projects (or indeed any project). Even large companies will generally not invest in projects with internal rates of return of less than 30 per cent. A domestic energy consumer will (unconsciously) usually be looking for very much higher rates of return still.

The fact that energy consumers need much more rapid 'paybacks' than energy suppliers has important consequences for the balance between investments in energy supply and energy efficiency. The market is distorted. We end up with much more energy supply than is in the financial interests of the consumer and the economy as a whole. We also end up with much more pollution into the bargain!

However, there is nothing inevitable about this situation. As we shall see in Chapter 7, there are various different strategems for ensuring a more even disposition of investments between energy supply and energy efficiency.

Let us now look at how the different levels of discount rates used by people and organisations are determined.

Levels of Discount Rates
The test discount rate used will vary tremendously between different organisations. However, energy supply projects (and also some energy efficiency projects organised by 'progressive'

energy utilities) will be funded by a mixture of funds from share-holders and banks. The rates of interest wanted by the banks and the levels of return wanted by the shareholders will be influenced by the general level of interest rates. These will shadow base interest rates. Base interest rates are set by the central banks on money lent to other banks, although individuals will have to pay rather higher rates if they want to borrow money from their local bank.

These interest rates include the effect of inflation. Thus a so-called real interest rate is calculated by subtracting the prevailing rate of inflation from the actual (nominal) base rate existing at any one time. As can be seen from Figure 5.3, there are significant differences between real interest rates prevailing in leading industrial states, the highest being in the UK. West German rates were also high in the early 1990s because of the effects of German reunification.

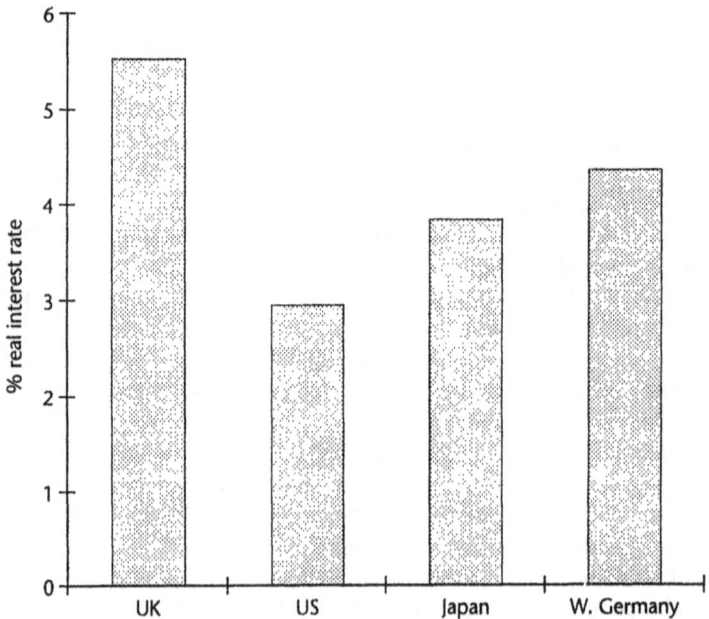

Figure 5.3 Average Real Interest Rates in Selected States, 1985–92

Source: *OECD Database* (London: Lloyds Bank Educational Service, 1992)

In practice even large, stable organisations like electricity utilities will have to borrow money at 1 to 2 per cent higher than the real interest rate (plus inflation). In addition shareholders will want rather higher interest rates than the bank to compensate for taxes and the risk factor (shareholders lose their money before the banks).

As a result, organisers of projects try to borrow the bulk of their money from banks rather than shareholders, although the banks will want to know that some people are prepared to risk their money if they are to have faith in it too.

The result of all these factors is that real internal rates of return for large conventional energy projects may be around 7 per cent in the US, but may be around 10 per cent in the UK.

The costs of capital intensive projects can be affected very greatly by the discount rate. The higher the discount rate, the higher will be the costs per kWh of a capital-intensive compared to a fuel-intensive (for example, fossil fuel) power plant.

Other Factors Affecting Capital Costs

Discount rates are not the only factors that can alter the capital costs of a given project; the assumed life of the plant is another. Energy utilities will often assess their own plant on the basis of its expected lifetime. However, sometimes, and usually in the case of independents, the project cannot be guaranteed a lifetime any longer than the length of their contract.

The supposed contract or lifetime length can make a big difference to a project. Contracts for supplying energy will usually be shorter than the expected life of the project. The shorter the contract, the higher will be the cost of energy produced.

The amount of time a plant is actually used is also important to the cost of energy from that plant. The quantity of energy produced by a plant in a given period compared to the plant's maximum theoretical output is called its capacity factor. (The quantity of energy produced by an energy system as a whole compared to its maximum output is called the load factor.) If a proposed plant is not going to be used very much then the capital costs are going to have to be low in the first place or else it will usually not be built.

Another factor affecting costs is the base year used. Inflation devalues the cost of money from year to year, so it is important to state a base year so that inflation effects can be taken into account.

Broadly similar rules apply to oil and gas extraction. Oil and gas companies will assess the returns on investments in new exploration, drilling, extraction and pipelines according to the rules I have already outlined. Existing fields will be abandoned when their running costs exceed the return.

Costs of energy from power projects are usually bandied about in newspapers without any mention of these criteria. However, if you are not told the discount rate, project lifetime, the capacity factor and the base year then the price quoted is of limited value. Often the costs of power from new energy projects are confused with the costs of power from existing energy projects whose capital costs have already been repaid or at least already committed.

Altering the Cost Criteria

Different ways of costing projects can produce wide variations in estimates of the price of electricity produced even when there is agreement on the running costs, fuel costs and the amount that the project actually costs to build. Take for example the Sizewell B nuclear power station. By 1993 some £2,700 million (before adding interest charges) had been spent on the project over more than seven years. Table 5.1 shows the range of prices using different discount rates and capacity factors. The estimate is done using 1991 prices and a 20-year lifetime, although these days power contracts are generally awarded for no longer than 15-year periods. (Use of a 30-year contract would reduce prices by around 10 per cent.) The bulk of the costs will be capital costs, the rest (fuel preparation, maintenance, allowance for decommissioning, and so on) have been calculated (some believe rather optimistically) to be about 1.5 p/kWh.

Table 5.1 Costs of Sizewell B Nuclear Power Station Using Different Criteria (1993 Prices)

Discount rate (%)	Capacity factor[1]	Price (p/kWh)
10	64	9.2
10	80	6.9
6	64	6.5
6	80	5.6

1. per cent of the time that the plant is operating at full capacity
Source: based on analysis and figures in Gordon MacKerron, 'The Economics of Nuclear Power', Science Policy Research Unit mimeo (University of Sussex, 1991)

In fact, MacKerron regarded the 64 per cent load factor as being the average for reactors of this type. On the other hand, Nuclear Electric, the developers, were more optimistic about the capacity factor and said that future, similar, models would cost less. (Chapter 11 looks further into arguments about the economics of nuclear power.)

The costs of electricity from fossil fuel power stations, whose costs are much more dependent on the fuel element than the capital cost element, are much less affected by changes in discount rates than capital-intensive power systems like nuclear power or, for that matter, wind power.

An example of the costs of power from a new combined cycle gas turbine (CCGT) is given in Table 5.2. This is now a proven technology which can, given the pipelines that go with such projects, usually achieve high levels of availability. An 85 per cent availability is assumed in Table 5.2 along with fuel prices of £2/GJ (about the level applicable in the early 1990s in the UK and the US), fuel to electricity conversion efficiencies of 50 per cent, capital costs of £500/kW, a 10 per cent discount rate and low running costs typical for CCGTs. A project lifetime of 15 years is assumed. Although the lifetime will probably be longer than this it is unlikely that the project will be able to negotiate a supply contract for natural gas at a given price for a period longer than 15 years.

Table 5.2 Cost Breakdown of CCGT Scheme

	p/kWh
Capital cost	0.9
Fuel cost	1.4
Operating cost	0.4
Total cost	2.7

At a 6 per cent discount rate the capital cost element falls to about 0.7 p/kWh. Since the other elements are not affected by the discount rate, the total cost at a 6 per cent discount rate is 2.5 p/kWh. As you can see this relatively fuel-intensive project is much less affected by changes in discount rates than capital-intensive projects like nuclear power or many types of renewable energy. The price of power from fossil fuel power stations is much more affected by changes in fuel prices.

In reality, once a power station has been built it invariably means that the cheapest path is to use the plant, however much it cost to build. Once the capital costs of building a project have already been incurred it is only worthwhile not to use the existing plant if the costs of another option are less than the fuel and running costs. These costs are called avoidable costs.

The avoidable costs of using Sizewell B nuclear power station should, in theory at least, be little more than 1 p/kWh and the avoided costs of the CCGT about 1.8 p/kWh. Hence it is much easier to argue on economic grounds against building new power stations than it is to argue for the closure of ones that have recently been completed. Of course, eventually, existing plant wears out, becomes unsafe or at least costs increasingly large amounts to maintain.

The final point that needs to be made is that the more use that is made of a particular resource (for example, natural gas), the higher will be the cost of supplying extra units of that resource.

The principles I have discussed for costing energy production projects can also be applied to energy-saving schemes.

Costing Negawatts

The notion of negawatts has been given widespread publicity by the US energy efficiency proponent Amory Lovins. He says that energy services can be provided by improving the energy efficiency of energy-using equipment.

We can calculate the costs of this energy efficiency in the same way as costing energy production by substituting the amount of energy saved for the amount of energy produced. One of the most popularised energy efficiency techniques has been that of energy-efficient lightbulbs. I can buy a 20 W electronic ballast compact fluorescent lamp (CFL) for £13.25 (1993 prices) at my local electrical shop. So the capital cost is £13.25. The CFL will give equivalent light to that which is produced by a conventional 100 W tungsten lightbulb. The fluorescent will save me 80 W when it is switched on, analogous to energy produced from a power plant. The fuel cost is zero and the running costs are negative since purchases of short-life tungsten replacements are avoided.

The capital cost of the CFL works out at 2.7 p/kWh at a 10 per cent discount rate. This assumes that the bulb shines for 8,000 hours and that the bulb is used for an average of five hours a day. The running costs are – (minus) 0.4 p/kWh producing a total cost of the electricity saved of around 2.3 p/kWh. The (unsubsidised) price of CFLs seems expensive compared to the running costs of ordinary lightbulbs only because domestic consumers will implicitly use discount rates of maybe 40 to 100 per cent compared to around 10 per cent used by power supply companies.

Often it is only cheap to promote energy efficiency through the replacement of machines at the end of their life-cycles. As is the case with other energy resources, the more energy efficiency is required, the higher will be the price of supplying the extra quantities of energy efficiency.

'Small' energy efficiency and renewable energy projects often have advantages over big power stations. Big projects may have economies of scale, but they lose heavily on the amount of interest that is charged on capital costs during construction time. Small energy projects (including most energy-saving schemes) can be bought off the shelf and installed very quickly, whereas big projects restrict utilities' room for manoeuvre in planning their systems and can cause problems if breakdowns occur. Energy efficiency and 'point of use' renewable energy systems will also avoid energy lost during transmission and distribution.

Having established a common set of criteria for assessing costs, I can now go on to look at the costs of the various options. Of course, just because an option is cheap does not mean it will be implemented. That depends on the access to capital and information enjoyed by the various actors, the way the market is regulated, cultural attitudes to different sorts of investment, the relative influence of (and advantages given to) various energy interest groups, and the nature of the technologies themselves.

We shall now look at the solution to environmental problems that has been the most widely applied in recent years: using more natural gas.

6

Gas, Gas and More Gas?

Use of natural gas is expanding. In the 15 years between 1976 and 1991 global natural gas use increased by 50 per cent, and its share of official world energy supplies rose from 19 to 23 per cent. Natural gas, given its availability, has become a first choice in increasing parts of the industrialised world for the provision of space and water heating. It is becoming a favourite supply of heat in industry and is also making forays into the electricity generation sector. Natural gas may even capture significant quantities of the fuel transportation market.

The technological basis for exploitation of the natural gas resource was established in the US in the 1930s as gas pipeline technology was developed. The first reason for the expansion of natural gas has been its technological advantages.[1]

Gas heating and, more recently, gas turbine technology is often cheaper to build, install and maintain than coal- and oil-based technology. The fact that gas is a clean fuel gives it a technological edge since the ease with which it can be handled often reduces the capital costs of the equipment used to burn the fuel. The equipment requires much less cleaning than coal or oil plant.

It is the economic advantage of natural gas technology that gives the fuel the edge rather than the cost of the raw material itself. (Coal is generally cheaper than natural gas as a raw material.)

Natural gas prices vary significantly between different countries. Higher prices exist in countries with few indigenous natural resources (see Table 6.1). Japan, which is not listed, has to pay very high prices for its natural gas imports.

In recent years, the development of gas turbine technology has allowed the fuel to become a more fearsome competitor in the electricity generation market. Gas turbines were used in the 1960s to provide 'peak' power because they were cheap to build, although they were more expensive to fuel in comparison with coal-fired steam turbines. However, gas turbines are becoming more and more efficient, a development which has been spurred on by the relentless drive for more efficient gas turbines to power jet airliners.

Table 6.1 1993 International Gas Prices[1]

	p/kWh
Canada	0.77
Australia	0.87
US	0.99
Netherlands	1.04
Belgium	1.13
UK	1.19
Sweden	1.30
France	1.39
Germany	1.43
Italy	1.95

1. Firm contracts with large users. Power stations will generally buy gas at cheaper rates than these, usually at about 0.7 p/kWh (approximately £2/GJ) in the US and the UK
Source: *Gas World International*, February 1994, p. 14

Since the 1980s a new generation of power station has been deployed called combined cycle gas turbines (CCGTs). These involve a high-efficiency gas turbine and a second, conventional, steam turbine similar to that used in conventional coal-fired power stations. The gas turbine sends exhaust gases out at a relatively high temperature (500 degrees C) that enables more electricity to be produced. This exhaust heat from the gas turbine can also be used in cogeneration systems to supply heating services. For a comparison between conventional steam turbines and CCGTs, see Figures 6.1 and 6.2.

Combined cycle gas turbines achieve higher rates of conversion of fuel energy input into electricity output (electrical efficiency) mainly because they achieve much higher fuel combustion temperatures than steam-raising systems. Steam turbine systems can increase steam temperatures only by increasing steam pressures. However, high steam pressures place great stress on materials, and little progress in achieving higher steam pressures (without incurring excessive costs) has been made since the 1960s.

CCGTs installed in the early 1990s convert around 50 per cent of the energy input into electricity compared to around 38 per cent for conventional 'clean' coal-fired power stations. The capital costs of CCGTs average at about £500/kW. CCGTs are, generally speaking, built cheaply and quickly compared to coal-fired power stations and they are also very cheap to operate. The conversion efficiencies of new CCGTs may reach 60 per cent in a few years' time. For a comparison of costs, see Table 6.2.

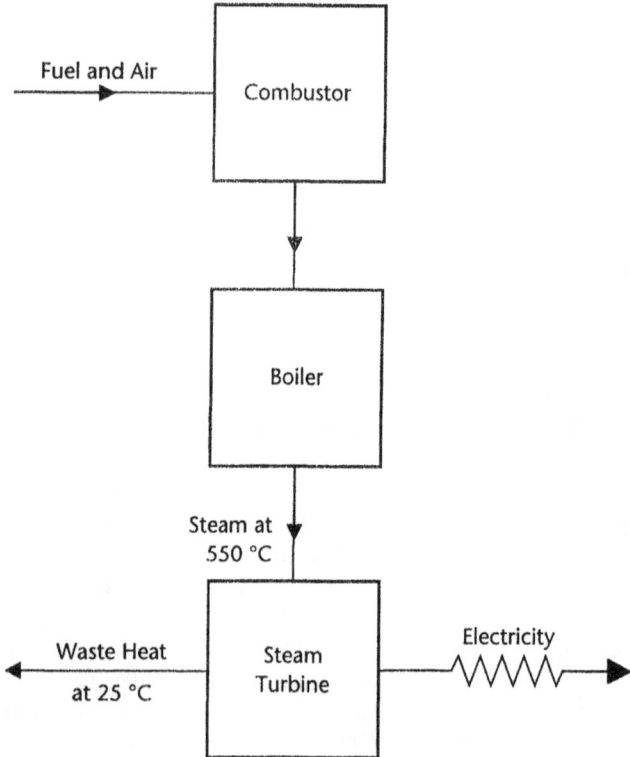

Figure 6.1 Conventional Steam Turbine

Note: Some data used in Figures 6.1 and 6.2 are drawn from D. Rooke, *Energy and Environment* (ed. B. Cartledge), Oxford University Press, 1992, pp. 45 and 49

The second reason for the expansion of natural gas has been the expanding known resource base. For example, from 1986 to 1992 global natural gas reserves increased by 35 per cent, almost double the rate of increase in natural gas consumption. Reserves in developing countries are being discovered at a rapid rate. Known Chinese natural gas reserves increased by 50 per cent between 1986 and 1992, for example.

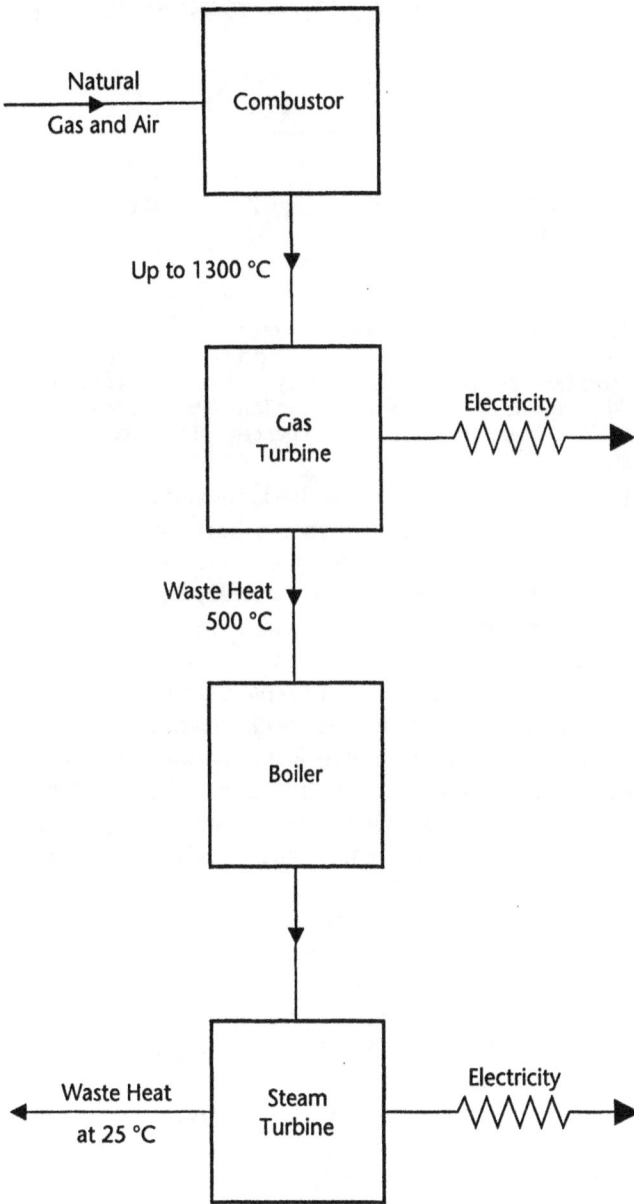

Figure 6.2 Combined Cycle Gas Turbine (CCGT)

Table 6.2 Natural Gas and Coal Power Costs (p/kWh)[1]

	Capital costs	Fuel costs	Operating costs	Total costs
CCGT (gas)	0.9	1.44	0.35	2.69
Conventional coal (with FGD and low-NOx burner)	1.35	1.2	0.6	3.15

1. These costs relate to new developments only
Assumptions: 10 per cent discount rate, 85 per cent capacity factor. *CCGT*: capital cost £500/kW, conversion efficiency 50 per cent, and natural gas cost of £2/GJ. *Conventional coal*: (with FGD and low-NOx burner) at capital cost of £750/kW, conversion efficiency of 38 per cent and coal cost of £1.30/GJ.

Natural gas resources in The Netherlands and in the British and Norwegian sections of the North Sea have lasted longer than initial estimates suggested. Of course, such resources are not everlasting; eventually production rates will peak. However, in the short term the resources may allow countries such as the UK, The Netherlands and the US, as well as other similarly gas-resource-blessed nations, the opportunity to expand further the proportion of their energy economies served by natural gas.

The third reason for the expansion of natural gas has been its environmental qualities. Indeed, those qualities which give gas an economic edge also give it an environmental edge. It has little sulphur and no ash (thus no particulate emissions) compared with coal and oil. Generally speaking, natural gas burns with much lower NOx emissions. Gas also has the benefit of producing only around 60 per cent of the carbon dioxide emitted when coal is burned to produce an equivalent quantity of energy.[2]

The expansion of natural gas to supply heating services in the UK during the 1960s was boosted by a national desire to cut down smogs caused by the emissions from coal burning. UK smoke emissions fell by a factor of four between 1952 and 1972.[3]

Developing countries are often stereotyped as tolerating appalling air pollution. However, as living standards increase and democratic rights spread there is increased pressure for pollution abatement. Natural gas is a big winner in these stakes. In Korea, for example, high cost Liquified Natural Gas (LNG) is being imported and its use being made compulsory in larger residential apartments.[4]

In Brazil the gas industry estimates that natural gas demand will more than triple between 1995 and 2005.[5]

In the US the 1990 Clean Air Act serves to give a boost to natural gas in the electricity sector since its effect is to make coal-fired power stations more expensive.

European targets for reduction of sulphur emissions from UK power stations and the UK's commitment to stabilise carbon dioxide emissions at 1990 levels by the year 2000 were used to justify a major programme of CCGT building in the UK.[6]

A comparison of the emissions produced by natural gas and other fuels in electricity generation is given in Table 6.3.

Table 6.3 Emissions from Different Electricity Fuel Cycles[1]

	SO_2 gS/GJ	NOx gN/GJ	CO_2 gC/GJ
Old coal (no controls)[2]	1,930	440	82,000
Conventional coal (new) (FGD + Low-NOx Burner)	170	230	72,000
Oil (no controls)	2,280	210	75,000
CCGT	67	74	34,000

1. On a fuel output basis
2. 1987 UK coal average
Source: N. J. Eyre, *Gaseous Emissions due to Electricity Fuel Cycles in the UK* (Harwell: ETSU, 1990)

The gas industry has tended to sell itself so strongly on the environmental front that the US oil industry has become alarmed. The *Oil & Gas Journal* was moved in a beginning-of-1993 editorial to attack the then incoming Clinton administration for its apparent preference for gas as opposed to oil. The editorial also attacked the priority given to several key environmental issues.[7]

Aside from the fears of competing energy supply interest groups there are environmental problems associated with an ever-increasing supply of natural gas.

The biggest environmental headache, apart from NOx emissions (which can be reduced with relatively cheap retrofits), is carbon dioxide emissions. It is true that gas produces only about 60 per cent of the carbon dioxide produced by coal in the course of supplying similar quantities of energy, and it is true gas technology often effectively reduces this proportion still further, yet large amounts of carbon dioxide are nevertheless generated. If the industrialised nations really have to reduce carbon dioxide emissions by 80 per cent then the problem is not going to be solved merely by substituting all coal and oil used by natural gas, especially as energy demand is increasing.

Carbon dioxide emissions are not the only contribution made by natural gas to global warming. Methane is itself a powerful greenhouse gas. Paul Crutzen has estimated that about 20 per cent of total methane emissions to the atmosphere come from fossil fuel sources. He says that three-quarters of this comes from natural gas use and the remaining quarter from coalmines.[8] It is probable that Western natural gas networks have low leakage rates, although no doubt more regular refurbishment of pipeline systems could curb leaks and improve safety. However, the leaks from Russian pipelines have caused great concern. (Not least to Russian helicopter personnel who look for leaks by firing flamethrowers at the pipelines to see if there is ignition!)

It is thought that a very high proportion of the natural gas pipeline leaks come from Russia, with some accounts saying that 15 per cent of Russian gas is leaked from the pipeline system.[9] Italian oil and gas interests have agreed to help refurbish parts of the pipeline system in exchange for gas supplies to Italy, but the parlous state of the Russian economy is hampering efforts to repair the bulk of the system.

The overuse of natural gas will not only have environmental disadvantages. There could also be price rises and security problems.

A Gas Resource Crisis?

As in the case of oil, there is no imminent global shortage of natural gas. However, again as in the case of oil, many leading industrialised nations are either rapidly approaching or increasingly exceeding the levels of gas consumption that can be supplied from their own domestic sources. Like oil dependence, natural gas dependence may become an increasing problem for most industrialised states. Many people believe this may lead to higher gas prices and greater energy security problems.[10]

Natural gas has many technical and economic advantages over both oil and coal at the point of use, but it is much more expensive to transport. Although natural gas can be compressed and transported by tanker as Liquified Natural Gas (LNG), this is an expensive process. Otherwise natural gas has to be transported by pipeline. Because natural gas has a much lower energy density than oil (meaning that a given volume of natural gas contains much less energy than the same volume of oil at normal pressures), the costs of the pipeline systems are higher.

While natural gas is cheap when the consumer is relatively close to the gas field, the more gas demand expands the farther and farther away (and thus more expensive) are the sources of supply.

That having been said, a number of countries, including the UK and the US, may enjoy moderate increases in gas supply without suffering major price increases. The UK has a gas supply surplus until the year 2000 because of some gas power station projects being cancelled. The US is increasing its imports of natural gas from Canada.

However, nobody knows how long North Sea gas production can continue to expand. Sooner or later production will peak. If US natural gas demand expands too steeply then the Canadians may start wondering whether they ought allow their natural gas reserves to be run down purely for the purpose of providing US industry with natural gas at low prices.

Many industrialised states have much poorer access to natural gas than the UK or the US. Central, Eastern and Southern Europe, for example, are short of resources. In the early 1990s key companies in Italy, Germany, Finland and Greece signed contracts for increased natural gas supplies from Russia. Japan has to pay a high price to import LNG.

Some extra European gas demand can be met from increased production from the Norwegian and maybe British sectors of the North Sea and from The Netherlands, although extra supplies from Norway may be expensive because of the need to construct new pipelines.

If demand increases rapidly then more countries will become more reliant on politically unstable countries such as Russia and Algeria. Natural gas supplies are, technically, rather more vulnerable to disruption compared with oil supplies because of the much greater reliance on pipelines. Not only can gas fields in politically volatile regions stop production, but the pipelines, which often cross several national borders, can be cut off. Gas-consuming countries cannot switch to a different supplier as easily as they can in the case of oil – not that there was anything 'easy' about the oil crises of the 1970s!

Asia and many parts of Africa are increasing their oil dependence on the Middle East. The same pattern will emerge in the case of natural gas, and security problems will increase.

Thus it can be seen that there could be both environmental and resource problems if natural gas use expands too far. There could be 'gas resource crises' that could in some ways be even worse than the oil crises of the 1970s and 1980s. Such crises would put up world energy prices in general since fossil fuels act, to a certain extent, as substitutes for one another.

The best use of natural gas, in environmental terms, is to substitute for coal and oil in the context of declining overall energy consumption in leading industrialised nations and to fuel no more than a moderate expansion of energy consumption in the rest of the world. Natural gas resource problems will certainly emerge all the more quickly if overall energy consumption is allowed to carry on growing at historic rates and natural gas consumption is rapidly pulled upwards to accommodate such growth. If natural gas is used very efficiently then both environmental and economic problems may be overcome.

7

Energy Efficiency

Using energy efficiently, that is using less energy to produce a given level of service, can have a number of economic as well as environmental benefits. Nations can be less dependent on foreign energy supplies. Energy prices can be held down. Energy efficiency is often a cheaper method of supplying specific energy services than buying in more energy. Energy efficiency is a very clean source of energy services. It will combat acid rain, smog, global warming and other environmental problems associated with energy use.

In the past, energy conservation has been associated with forced savings made during periods of high oil prices. Nevertheless, as has been suggested in Chapter 5, energy efficiency can be used to deliver energy services more cheaply than merely supplying energy if the market barriers to energy efficiency can be overcome.

This chapter examines some of the technologies and policies involved in delivering energy efficiency.

Supply-side Energy Efficiency

The bulk of electricity is produced by burning fossil fuels. The resultant heat drives a turbine which rotates a magnet around a coil of wire. An electric voltage is induced in the wire. This process involves immense wastage of energy because of the consequences of the second law of thermodynamics which effectively limits the degree to which heat energy can be converted into mechanical energy.

This explains why even the most advanced electricity-only power stations in commercial operation (CCGTs) waste around half of the energy input. The waste heat goes up cooling towers. This waste is shown in Figure 7.1. Power stations built in the early 1970s waste at least two-thirds of the energy input.

There are ways of reducing the energy wasted by electricity-generating systems, notably cogeneration (known in the UK as combined heat and power) which involves producing electricity and heating services at the same time.

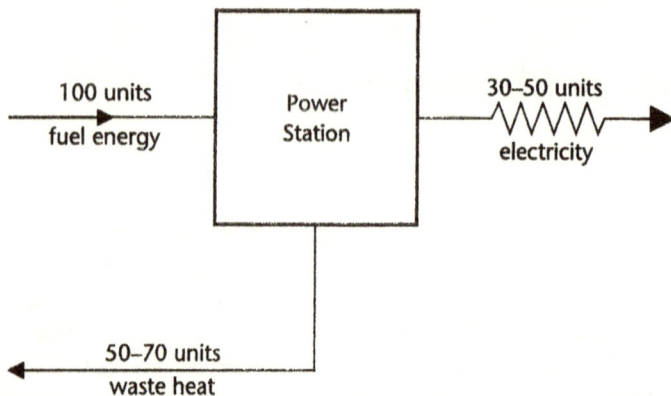

Figure 7.1 'Electricity-Only' Power Station

Cogeneration

A cogeneration system can turn much larger proportions of energy input into useful energy than can electricity-only plant.

Cogeneration can supply both electricity and space- and water-heating services to industry, commerce and homes. In properly organised schemes the cogeneration plant will be sited close to the heat demand (heatload).

A gas-fired cogeneration system, with typical efficiencies for plant, large numbers of which are being installed in The Netherlands and Denmark, is shown in Figure 7.2.

As can be seen, only a small portion of the energy is wasted. The system turns over 80 per cent of energy input into useful energy called the overall energy efficiency.

These days the bulk of new cogeneration systems are powered by natural gas. Sometimes cogeneration units will be co-fired with oil. Coal, wood or even rubbish can also be used as fuel.

Technological factors are giving gas-fired cogeneration systems a big advantage over coal-fired systems. The difference is that the temperature of the gas turbine exhausts is much higher for a given level of electrical efficiency than the temperature of the water that is discarded by steam turbine systems. The heat of the gas turbine exhausts can be used to provide either more electricity, or heating services for a variety of uses. The tepid water that exits from steam-raising systems is practically useless. Gas engines can have much the same advantages as gas turbines in the case of

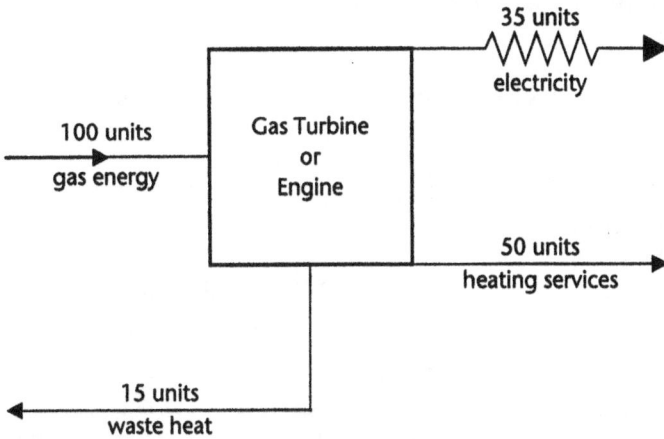

Figure 7.2 Gas Cogeneration

smaller projects. Steam-raising systems can be, and in the past have, sometimes been used as part of cogeneration systems. However, steam-based cogeneration systems are usually less energy efficient than gas-based systems.

Aside from this the maintenance and capital costs of gas-fired cogeneration systems are low compared with coal-fired units, especially in the case of smaller systems. Cogeneration is often most economically implemented in relatively small packets. The efficiencies of these smaller systems are increasing and their costs are falling.

Although coal-burning systems have been devised that can use gas turbines by gasifying the coal first (see Chapter 9), such systems involve much higher capital costs than natural-gas-fired systems and they can be deployed only in quite large sizes.

The Costs of Cogeneration

Cogeneration systems need to be installed in the right circumstances. This means having an appropriately sized heating need (heatload) to which to send heat, having the ability to send surplus power to the electricity grid for reasonable prices and the likelihood of a high capacity factor (it must be used for at least 50 per cent of the time as a rule of thumb). Well-sited cogeneration systems can produce electricity for as little as 2 p/kWh. Large quantities of cogeneration resources exist in the 2–3 p/kWh cost range (see Table 7.1).

Table 7.1 Cost Structure of Gas Cogeneration System

	p/kWh	
Capital cost	1.0	
Operation and maintenance	0.45	
Fuel costs	0.68	(3.09–2.42 credit for heat services supplied)
Total cost	2.13	

Assumptions: 10 per cent discount rate, 15-year project lifetime, 34 per cent of energy input turned into electricity, 51 per cent turned into heat of heating services. Fuel cost £3/GJ (1993 prices), assumed efficiency of heating system displaced by cogeneration system 65 per cent, capacity factor 75 per cent.

Gas engine and gas turbine cogeneration technology is, kW for kW, as cheap as a large CCGT power station to install, provided the pipework (needed for district heating systems) is not too extensive.

The economic size of cogeneration systems is declining as the capital costs of small systems decline. Machines as small as 40 kW are now available. There are experimental engines as small as 5 kW (the average household consumes 1.5 to 2 kW of electricity).

Gas turbines are generally used for schemes of over 1 MW in size. The economic size of gas turbine technology is falling.

Fuel Savings and Pollution Abatement

Cogeneration can currently turn around only 30–40 per cent of the energy input into electricity output. This electricity conversion efficiency is improving along with improvements in gas turbines. Most of the energy that is not converted to electricity can be provided as heat. This means that the overall energy efficiency, or the proportion of energy input that is turned into useful energy output, can be (and is in some systems) close to 100 per cent.

Table 7.2 shows how three different energy systems produce the same amount of electricity and heat (coal-electricity plus separate gas heating; gas cogeneration; and CCGT gas-electricity plus separate gas heating). They use different quantities of energy and produce different amounts of carbon dioxide emissions.

Table 7.2 Cogeneration and Electricity Production Compared for Fuel Use and Carbon Dioxide Emissions

	Energy produced (PJ)	Gas used (PJ)	Total energy input (PJ)	Carbon dioxide (tonnes)
Traditional coal-fired electricity	102 electricity 154 heat	237	537	48,220
Natural gas cogeneration	102 electricity 154 heat	300	300	15,180
Combined cycle gas turbine (CCGT)	102 electricity 154 heat	441	441	22,320

Notes: the old coal-fired power stations are likely to be the first to be retired, and therefore the cogeneration systems will substitute for them. The figures for gas cogeneration are based on systems deployed by Isselmij energy company, a municipally owned electricity/gas utility in The Netherlands. Their systems, installed in the 185 kW to 850 kW range, turn 34 per cent of the gas energy input into electricity and send out 51 per cent of the energy as heat, giving a total overall energy efficiency of 85 per cent.

In the cases of coal-fired electricity and CCGTs, old gas boilers (which supply heat) are assumed with efficiencies of 65 per cent.

The figures in Table 7.2 suggest that if gas cogeneration is substituted for coal-fired power stations and gas heating supplied separately energy consumption is reduced by 44 per cent and carbon dioxide emissions are reduced by around 69 per cent. CCGT electricity-only production also reduces energy consumption and carbon dioxide production, but by only 18 and 54 per cent respectively.

SO_2 and NOx emissions will be reduced by around or over 90 per cent by the gas systems compared with the coal-fired power station. Selective catalytic reduction units can be added to cogeneration units further to reduce NOx emissions, albeit at moderate extra cost.

Although switching from coal-fired electricity production to gas cogeneration increases gas use, the increase is much smaller than with using CCGTs. The remaining increase could be offset by other energy efficiency improvements.

Cogeneration seems to have overwhelming advantages, so how come it is not more widespread already and how can its development be promoted?

Development of Cogeneration

In the past cogeneration has always been restricted in its application by the structure of energy markets. Cogeneration is best organised when an efficient cogeneration machine is matched to a heatload, say a factory or a housing estate.

Yet traditional electricity suppliers sell electricity, not heat, so they are unlikely to place priority on installing cogeneration plant. What institutions exist to supply cogenerated services in most countries? Answer, none. Thus it often does not happen, despite many very cost-effective opportunities. Even where industrial companies do have great possibilities for cogeneration, many do not have good access to capital, and they have to pay high prices for natural gas supplies. Moreover, the cogeneration projects that do exist often obtain poor rates for the electricity that they 'export' to the grid.

If, as described later in this chapter, rules are changed to ensure that the potential for cogeneration is tapped, then the energy suppliers can use their own access to capital and can negotiate cheap natural gas prices by organising several schemes at once. This is being done by the energy companies in The Netherlands, for example.

Cogeneration in Germany
In Germany the market for cogeneration has, in the last few years, expanded rapidly in the shape of small-scale gas cogeneration using engines. This type of technology is becoming cheaper and is being adopted widely in industry. In 1990 cogeneration supplied around 8 per cent of German electricity. This proportion is expanding. The German government has set aside considerable sums to enable the existing district heating systems in former East Germany to be linked to cogeneration systems.

However, the Germans have as yet to ensure a thoroughgoing cogeneration programme promoted by interventionist means. Partly this results from German faith in market mechanisms.

Danish and Dutch Cogeneration
In Denmark and The Netherlands the systems being installed have overall energy efficiencies of over 80 per cent, and in recent Danish cases as high as 95 per cent.[1]

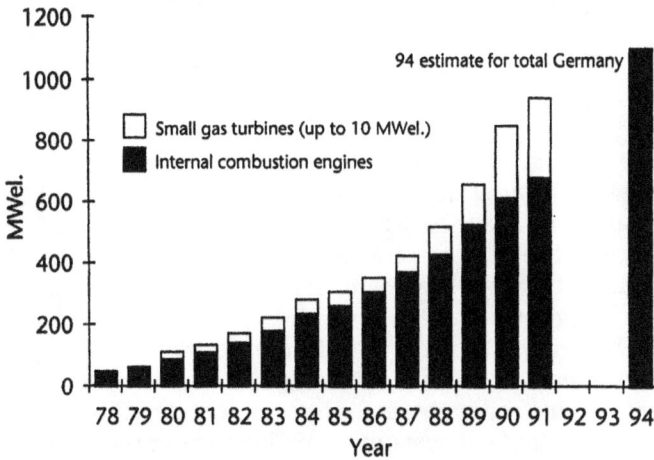

Figure 7.3 Development of Small Cogeneration
Capacity in West Germany

Source: Ruhrgas, HMUB/IWU

Denmark leads the world in the field of cogeneration. By the end of the 1980s around 38 per cent of all electricity generated in Denmark was coming from cogeneration plant.[2] The Danes have built up a district heating system over the last 20 to 30 years which has increasingly been fed with heat from cogeneration units. Over a third of district heating is now supplied by cogeneration. Expansion of their cogeneration system (and using natural gas in place of coal) will be the most important means of achieving their plans to reduce 1988 levels of carbon dioxide emissions (outside the transport sector) by 20 per cent by 2005.

In the late 1980s the Danes gained access to North Sea gas and new cogeneration units are now being fired with, and old coal-fired sets converted to run on, natural gas.

Since 1990 many cooperative cogeneration schemes have been formed by local residents to supply heating to, say, housing estates of maybe 300–400 houses. These tend to have especially high overall energy efficiencies.

The Dutch, who are rapidly expanding cogeneration, are demonstrating that the technology can be cost-effectively and quickly implemented even when when natural gas already supplies most heating needs at low prices. High overall energy efficiency gas cogeneration will conserve Dutch gas reserves and prevent the country becoming dependent on expensive imports of natural gas.

Gas cogeneration systems are being installed in industry, commerce and on new housing estates. The latter send heat through district heating systems and these are earmarked for linkage to larger gas turbines some time in the future. Many schemes are being established by the municipal energy distribution companies and these have overall energy efficiencies of over 80 per cent. These systems are set up to match the heatloads with the electricity going straight to the grid. This ensures a high overall energy efficiency.

In 1990 around 15 per cent of Dutch electricity was supplied from cogeneration that is mostly sited in industry, especially the chemical industry and the large greenhouse sector. It is expected that cogeneration will expand to produce 30 per cent of Dutch electricity soon after the year 2000.

Cogeneration is expanding rapidly in other countries as well: for example, Finland already has 30 per cent of its electricity produced in this way.

Cogeneration in the US
In the US cogeneration has been advanced in recent years by the non-utility generators at a rapid rate. PURPA (Public Utility Regulatory Policies Act) rules, first introduced in 1978, insist that electricity utilities offer good terms to independent renewable and cogeneration schemes.

The number of independent schemes has been growing, especially in recent years. If plans made by the middle of 1993 are carried out, the proportion of electricity supplied by non-utility generators (NUGs) would increase from 8 per cent in 1990 to roughly 20 per cent in the year 2000.

Under the terms of the 1992 Energy Policy Act the Federal Energy Regulation Commission now has the authority to order utilities to afford NUGs 'point to point access' to utility transmission systems. This effectively ends the utilities' monopoly over generating electricity. In some states, including California, the NUGs now supply all new (fossil fuel) generating capacity and contracts for tranches of cogeneration capacity are allocated through a competitive bidding system.

Over the next few years deregulation (abolishing the utilities' monopoly to supply power to consumers) will allow so-called 'retail wheeling', so allowing NUGs direct access to the energy users. However, there are doubts as to whether this will always help cogenerators since the guaranteed market shares currently accorded to cogenerators in some states may be abolished as part of the move towards deregulation.

The numbers wanting to move into the cogeneration supply market are vast. In California, for example, cogeneration proposals are especially strong in industries like steel, oil refining and enhanced recovery oil drilling as well as in the food industry. In 1993 one company, Southern California Edison (SCE), received 11,000 MW of bids from independents, mainly industrial cogenerators, but only 600 MW of contracts were actually awarded. The cogenerators are sometimes paid no more than 3 c/kWh, and very rarely more than 4 c/kWh, by utilities. Meanwhile 5,000 MW of mostly conventional electricity power plants were supplying electricity to utilities in California for upwards of 10 c/kWh under contracts dating from the early to mid-1980s.[3]

Despite the growth of cogeneration in the US, the definition of a cogeneration project is, under PURPA rules, very lax. This has meant that many schemes have low overall energy efficiencies. In some circumstances a scheme with less than 50 per cent overall energy efficiency can enter the market under the banner of cogeneration. The PURPA was originally concerned as much with promoting the growth of an independent power sector as with reductions in polluting emissions.

UK Cogeneration Prospects
There is a concentration on electricity-only plant in the UK. Cogeneration receives little support from either the main generators (PowerGen and National Power) or the regional electricity companies (RECs). Some cogeneration schemes, in particular the giant 1,875 MW plant on Teesside, are similar to many US cogeneration facilities in that they have a relatively low thermal efficiency (under 60 per cent). Deregulation of the large industrial electricity supply market has, since 1990, failed to produce a big increase in cogeneration, although there is a steady trickle of small schemes being set up.

In 1990 less than 3 per cent of British electricity was supplied from cogeneration plant. The UK government projects a big expansion in cogeneration projects so that maybe 15 per cent or more of UK electricity is supplied from cogeneration by the year 2000. There are as yet few incentives in place to achieve this target and no means to ensure that the cogeneration schemes have high overall energy efficiencies.

Denmark, The Netherlands and Finland are very much the exceptions as far as promotion of cogeneration is concerned. Leading industrial countries such as Germany, Japan, the UK and (as far as overall energy efficiency levels are concerned), the US, lag far behind. In Japan it has actually been illegal for independent electricity producers to sell electricity which is in excess of their own requirements. Although cogeneration is generally expanding in many places, sole reliance on market forces will not achieve the technology's full economic potential.

Given the progress made in places like Denmark and The Netherlands, and with smaller and smaller cogeneration systems becoming economically viable, high overall energy efficiency cogeneration could economically supply around 50 per cent of electricity requirements in a country like the UK.[4]

This amount of gas cogeneration could, if used in place of coal, lower electricity prices, reduce national carbon dioxide emissions by nearly 15 per cent, and use a great deal less natural gas than would be used by building CCGT plant instead.

Fuel Cells

Cogeneration is not the only technological development that can improve supply-side efficiency. As was mentioned in Chapter 6, the conversion efficiencies of CCGTs are slowly moving upwards although progress is inevitably limited by the laws of physics. However, there are some technologies which avoid some of the problems involved in turbine technology. Fuel cells are showing great promise.

Fuel cells convert energy directly into electricity through electrochemical means. Fuel cells are like batteries, except that the fuel (for example, natural gas or possibly hydrogen) is constantly being replenished. Some small fuel cells have already been deployed, for example to run buses in Vancouver. The cost is still high, but according to the Electric Power Research Institute, 2 MW 'molten carbonate' fuel cells with efficiencies of up to 60 per cent could be in commercial use by the year 2000. In volume production the capital costs could be competitive with those of CCGTs.

European researchers working on solid oxide fuel cells hope that efficiencies of over 65 per cent could be achieved.[5]

Fuel cells will produce much lower NOx emissions than even CCGTs with de-NOx equipment. Thus they are attracting interest from environmentalists.

Another technology that could improve the efficiency with which electricity can be produced is magnetohydrodynamics (MHD). This has similarities with conventional power stations in that the fuel is burned, but instead of driving a turbine the heat energy is passed directly through a magnetic field to produce an electric voltage. However, this technology is still in an under-developed state.

Fuel cells and MHD systems could also be incorporated into cogeneration systems.

Of course the potential for cogeneration, although massive, is inevitably limited by the amount of energy wasted during the course of electricity production. End-use efficiency is ultimately limited only by the size of energy demand as a whole.

End-use Energy Efficiency

In 1975 American physicists Marc Ross and Robert Williams calculated that energy equipment in use usually achieved less than 1 per cent of its theoretical efficiency.[6]

Indeed, searching for the essential energy use is rather like peeling an onion. As we have seen in the previous section, much energy is wasted producing electricity. Yet when analysis of the use of energy is done, further layers of energy waste peel off. Eventually we are left with a very small core and a pungent onion-smell analogous to the smell of pollution arising from energy usage.

A lot of energy is lost during transmission and distribution (T&D) to the consumer. Natural gas pipelines consume energy because they need to be pressurised. Electricity losses occur in T&D because of electrical resistance. These losses are around 8–10 per cent of total electricity produced in developed countries, and much higher losses occur in many poor countries. In Bangladesh around 40 per cent of electricity was lost in T&D during the 1980s.[7]

When one actually arrives at the end-uses, for example use of electricity in industry, the onion-peeling process accelerates.

In manufacturing, roughly half of all electricity is used to power motors. The motors are themselves 80 to 90 per cent efficient, although improvements in efficiencies of 5–10 per cent can be cost-effectively made. These motors drive pumps or fans, and here there are major losses in efficiency. This is because conventional pumps either work at full power or not at all, and not according to the rate at which the liquid or gas being pumped actually needs to be circulated.

Variable speed drives can be fitted which can slow down the motor drives according to need. This can be done through retrofits. Investments in variable speed drives can often pay back in months rather than years with an electricity-saving cost of 1 p/kWh or less (10 per cent discount rate).[8]

The onion-peeling does not stop here. The amount and speed of gas or liquid that needs to be pumped can be reduced with better-designed systems and new types of heating methods. Computerised energy management systems are becoming cheaper and more effective. Similar exercises can be performed in virtually all areas of energy use.

What Energy Is Used For

Table 7.3 Proportions of National Energy Use by Sector in OECD Countries

	%
Transportation	20–30
Manufacturing	30–50
Services/commercial	10–20
Residential	20–30

Note: these figures (also in Table 7.4) have been adapted to include allowances for energy lost during electricity production
Sources: Energy policies of IEA countries (*1991 Review*), US DOE, UK DTI and assorted studies

Table 7.4 Proportions of Energy Use According to Application in OECD Countries

	%
Transportation	20–30
Space heating	20–30
Air-conditioning and ventilation	0–10
Process heat	15–25
Motive power	8–15
Appliances	5–10
Lighting	5–10
Water heating	5–10

Sources: see Table 7.3

Electricity production typically accounts for roughly 35 per cent of total primary energy use in OECD countries.

Energy use in OECD countries is expanding most rapidly in the transportation and service sectors, with use of electricity expanding faster than any other form of delivered energy. A brief look at some of the possibilities for energy efficiency measures follows.

Transportation

Because of our ever-increasing love for the motor car and the relatively slow improvements in motor vehicle efficiency over recent years, transportation is the most difficult sector of energy consumption to control. Recent low oil prices have reduced the pressure for policies that will encourage motor car fuel efficiency.

In spite of this, some people are predicting a technological leap forward in motor vehicle production involving the development of 'superlight' and thus very energy-efficient vehicles (see Chapter 8).

Air transport is expanding even faster than transport by motor cars, although improvements in aircraft efficiency are reducing the impact of this increase on fuel demand. World air travel quadrupled in the 1970 to 1990 period and aircraft fuel efficiency more than doubled. Around 15 per cent of transport fuel burned by the UK and US is used in aircraft.

Buildings

Approaching half of all energy used in the industrialised world is concerned with maintaining the internal environment of buildings: space heating, air-conditioning, ventilation, lighting and water heating.

Space heating needs can be dramatically reduced by insulating houses properly when they are built. Building regulations have become more strict about energy efficiency than they used to be. Thus, a house built 100 years ago will, without extra insulation, consume around four times as much energy for space heating than a house built to contemporary specifications.

A new Swedish or a Swiss building will consume half as much as a new British building. Thicker wall and roof insulation are the main differences, together with techniques such as fitting triple glazing which incorporates specially coated 'low emissivity' glass to stop the outflow of heat. There are houses in existence that need practically no energy at all to run them.[9]

Lax building regulations are a surefire way of losing money for whoever will be living in them, whether in ten months, ten years or 100 years. It does not cost that much extra to build energy-efficient houses. The trouble is that the people who build them do not have to live in them.

Energy efficiency measures can reduce air-conditioning costs. Cheaply manufactured 'low emissivity' glass can cut down on a building's heat gain and allow smaller (and cheaper) air-conditioning systems to be installed. A study of commercial buildings in Bangkok, Thailand, conducted by the Lawrence Berkeley Laboratory, concluded that one plant manufacturing enough low emissivity glass panels to eliminate the need for a billion-dollar power plant could be established for just $10 million.[10]

The relatively well-insulated state of new housing and a shift towards natural gas heating (as opposed to energy-wasteful electric heating)[11] means that the US Department of Energy (DOE) is predicting a decline in carbon dioxide emissions associated with the US residential sector. The use of energy-efficient gas-condensing boilers will enhance this trend.

Building environments are increasingly controlled by microchips which turn the heating and air-conditioning up or down according to fine-tuned calculations, and switch the lights on or off according to whether there is anyone around. Energy use in buildings is increasingly being controlled by off-site computer monitoring techniques.

Air-conditioning systems will, in the future, save energy by depending more on use of gravity and less on pumps, and use systems involving night generation and storage of ice. A growing body of opinion says that near-zero-energy air-conditioning can be achieved by improvements in building design. Buildings are being constructed that mimic the cooling methods used by termites to keep their mounds cool. Ventilation systems are controlled by a Building Energy Management System.[12]

The amount of electricity needed for lighting could plummet in the future. Presence detectors will switch off lights when nobody is around and lighting systems which deliver more light for less energy use will be deployed. You may have heard of compact fluorescent lamps, but you may not be familiar with the special reflectors that can be fitted to office luminaires. They can give more light with half (or less) the number of fluorescent strips. The cost of saving electricity through buying and fitting reflectors is under 2.5 p/kWh in the UK[13] and the cost is falling fast while the efficiency of the units is increasing. The amount of electricity used for lighting purposes can be reduced, in many circumstances by a quarter, merely by replacing the solid ballasts in fluorescent units with electronic ballasts.

It has been estimated that if just 30 per cent of conventional tungsten lightbulbs being used in the former Soviet Union were replaced by compact fluorescents then electricity savings would be equivalent to the output of ten Chernobyl-type nuclear power stations.[14]

An example of the potential for energy efficiency in the US lighting sector is given in Table 7.5.

Table 7.5 Lighting Efficiency in the Commercial Sector Over 20 Years

---

Assumptions:

1. An aggressive energy efficiency programme in all US states.
2. 1990 US lighting commercial electricity demand: 1.14 quads.
3. Forty-two per cent increase in floorspace 1990–2010 (US DOE projection).
3. Assume that all new future fittings reduce electricity/lighting needs by an average of 50 per cent compared to traditional techniques by a combination of new fittings and other lighting control techniques.[1]
4. Assume that 80 per cent[2] of fittings existing in 1990 are replaced by new fittings by 2010 resulting in a 50 per cent reduction in power needs by 2010.

A reduction of 18 per cent in electricity consumption for lighting purposes results. US DOE projects an increase of 11 per cent.

---

1. The 50 per cent figure will, in a few years, be conservative considering the rate of advance in light fitting and lighting control techniques
2. This target is not unreasonable considering the declining costs of the fittings and considering that other lighting efficiency measures installed in 'new build' offices will also reduce power demand

The turnover of buildings is slow, but a great deal can be done to retrofit existing buildings with insulation. Moreover, a range of devices that will improve energy efficiency, many of them computer controlled, can be installed for costs that are less than the costs of power from new power stations or the cost of extra gas through new pipelines.

Generally, the energy consumption of uninsulated old houses can be reduced through insulation by roughly 20 per cent with a pay-back period of around three years. The amount of energy used by heating boilers can usually be cut by around 30 per cent by replacing old boilers with natural gas condensing boilers.

Manufacturing
Manufacturing processes are highly differentiated, but there are many common energy efficiency techniques that can be applied. I have already mentioned more efficient motors and variable speed drives.

Besides the use of electricity to drive motors, process heat is the other principal energy use in industry. Over the past 20 to 30 years there has been rapid development of waste heat recovery techniques such as heat exchangers and, most recently, regenerative burners. These incorporate the heat exchanging function within the burner itself.

Much of industry would be better served by contact heaters rather than the traditional but less energy-efficient air-blown techniques.

Many companies have inefficient boilers that are several decades old. They could be replaced with much more efficient systems, but the companies do not have the money to do so.

Of course, manufacturing industry, with its large heatloads, is often ideal for cogeneration, although use of cogeneration should also be combined with the maximisation of end-use efficiency.

Computers are becoming increasingly important in improving energy efficiency through computer-aided design, which can help to design energy-efficient systems, and through the monitoring and control of energy use. Microprocessors can control energy used by machines and systems so as to ensure, for example, that energy is not used when the machines are not running on full load. Computer systems can study the interaction of machines involved in fabrication of products and reduce energy consumption to an optimised minimum.

It is likely that there will be a shift towards the production of products and services that involve less energy to manufacture and which consume less energy when used compared to traditional types of industrial products. This will be discussed further towards the end of this chapter.

An increasing number of companies now adopt energy efficiency targets as important parts of their 'green' corporate plans. IBM, for example, has achieved targets of 4 per cent annual reductions in energy use since the late 1980s.

Commercial/Service Sector
This sector is expanding much faster than traditional manufacturing, yet its energy consumption per unit of output is a great

deal smaller. Indeed, a very high proportion of the energy costs are building costs. The main types of building energy efficiency techniques have already been described. Apart from energy used to maintain building environments other uses of energy in the commercial sector mainly involve appliances like computers, fax machines and photocopiers.

Computers are the fastest rising energy-eaters of the service sector. They made up 5 per cent of the US commercial sector's electricity consumption in 1990, a proportion which may double by the year 2000. Much of the computer-guzzled power ends up as waste heat which in turn makes air-conditioning systems use more energy.

A range of energy-saving techniques is likely to be incorporated into computers in the future, partly prompted by the US EPA's programme of awarding computers which meet energy efficiency targets with an 'energy star' rating. The US government has concentrated computer manufacturers' minds by launching a policy of buying only energy-efficient computers. Energy-saving computer techniques include avoiding the need for fans to clear away waste heat by using 'heat sinks', making computers power down when not in use or using chips that go into sleep-mode in between keystrokes.

Energy-gulping cathode-ray screens and disk drives are likely, in the future, to be replaced, respectively, with liquid crystal displays and so-called flash EPROM memory cards. Laptop computers which need to conserve energy because of the constraints of batteries already use much less power than desktop models, and companies like IBM are already beginning to market 'green' desktops that use only a small fraction of the power used by early 1990s models.[15] The market for desktops is rapidly expanding and this will herald a market opportunity for solar power as a replacement for costly batteries (see Chapter 12).

The US Electric Power Research Institute (EPRI) says that by redesigning photocopiers some 70 per cent of their energy use could be cut.[16]

Office information machines are in the process of being integrated and this development is likely to improve energy efficiency.

Residential sector
As is the case with the commercial sector, the bulk of energy used in homes is concerned with lighting, heating and air-conditioning the interior of buildings. However, the energy consumed by appliances such as fridges and freezers amounts to about a third

of domestic energy use. The energy efficiency of these appliances can be dramatically, and cost-effectively, improved. Door seals that last longer, better insulation, and various improvements in the efficiency of the cooling systems can cut the energy costs of fridges and freezers.

In fact there is a big overlap between the costs of energy-efficient and inefficient machines; the cost of saving energy by buying a fridge/freezer that consumes less than two-thirds of the energy of a similar-sized average machine has been found to be around 2.2 cents/kWh (7 per cent discount rate).[17] A similar story can be discovered in the case of most, if not all other, domestic appliances ranging from dishwashers to showerheads.

The efficiencies of 'state of the art' electrical appliances being installed in a German low-energy demonstration house in Freiburg are compared with appliances with average energy efficiencies in Table 7.6.

Table 7.6 Energy-efficient Equipment Planned for Ultra-energy-efficient Demonstration House

End use	Annual electricity consumption (kWh)	
	This house	Conventional house
Lighting	90	380
Refrigerator	110	530
Freezer	110	780
Washing-machine	150	380
Dishwasher	60	380
Ventilation	60	–
Television	30	220
Other appliances	150	600
Total	760	3,270

Note: ventilation is necessary for low energy houses in order to provide air circulation
Source: D. Olivier, *Energy Efficiency and Renewables: Recent Experience on Mainland Europe* (Credenhill, Herefordshire: Energy Advisory Associates, 1992), p. 34

Regardless of energy efficiency programmes, the energy used for cooking purposes is being reduced by the introduction of microwave ovens. Indeed, the 'saturation' of demand for many appliances such as refrigerators means that improvements in energy efficiency of such machines will result in absolute reductions of energy use. The fact that many sectors of demand for energy services have become saturated is one of the reasons why grandiose projections of future energy consumption in leading industrialised countries made by energy supply-oriented analysts have been confounded in recent years.

Future growth in demand for consumer products is likely to be met more by providing higher quality services tailored more to satisfying individual tastes rather than churning out ever-increasing numbers of standardised versions of energy-intensive heavy metal objects.

Thus it can be seen that there is a massive potential for energy efficiency savings in the main sectors of the economy. Just how big are the savings that can be realistically and cost-effectively implemented?

How Much Energy Efficiency?

There are many studies projecting large energy savings and reduced energy bills.

In the early 1980s a British study estimated (on the basis of average rates of economic growth) that UK energy consumption could cost-effectively be reduced by almost a half by 2025.[18] Since then the 'state of the art' energy efficiency equipment has advanced considerably, so that their study, if done today, would project even bigger savings!

In the US, a 1989 survey published by ACE[3] based on modest assumptions projected an 11 per cent decline in 1987 carbon dioxide emissions by the year 2000 if a clutch of cost-saving measures was introduced.[19]

In West Germany the Enquete Commission, whose findings were accepted by the West German government, found that energy efficiency measures could reduce 1987 levels of West German carbon dioxide emissions by over 23 per cent by 2005, and still be profitable.[20]

There is evidence to suggest that if awareness of and pressure for energy efficiency can be created, then the costs of energy efficiency will decline.

Declining Costs of Energy Efficiency

Currently, the market for energy-efficient products is relatively weak. However, there is practical evidence to suggest that the creation and sustenance of a market in energy services may drive down the cost of energy efficiency measures much faster than the rate at which the costs of supplying energy are likely to decline.

This can be seen in the case of compact fluorescent lightbulbs which can produce the same light as traditional tungsten bulbs while using only a fifth of the electricity. In 1993 the (unsubsidised) UK retail price of a 100 W CFL was roughly 40 per cent cheaper than it would have been in 1989.

The cost of computer and microprocessor-based energy efficiency techniques has fallen dramatically in recent years. Indeed, not long ago such systems did not even exist. Information technology has opened up tremendous potential for using energy more efficiently.

Market Barriers to Energy Efficiency

I have mentioned two major market barriers to energy efficiency in Chapter 5, namely the much longer pay-back periods that, in general, exist for energy supply as opposed to energy efficiency projects and the lack of consumer knowledge about energy efficiency techniques.

Other market barriers include the fact that many energy consumers simply do not have any money to spend on investments in anything, that institutions will often be biased towards investing in items connected to their central business rather than in energy efficiency, and that consumers do not have the time to find out about energy efficiency techniques.

On top of this there is a dominant cultural disposition to regard energy efficiency as an inadequate way of satisfying consumer needs compared to energy supply. This cultural bias is, at least in some quarters, being changed. However, efforts to change policies to overcome the market barriers to energy efficiency are held back by energy supply interest groups whose influence is extremely well entrenched. The energy supply interest groups are often supported by economists.

Economists say that the most efficient economic out-turn is created by maximising competition; yet if the market is controlled by companies which favour only one type of investment (supply-side investment) and who maximise their revenues by maximising energy sales, the market cannot deliver the most economically efficient (or for that matter environmentally desirable) result.

The investments in energy conservation made by consumers (even when assisted by government grants and tax incentives) are extremely small compared to the investment in supply-side capacity made by the main energy utilities and the oil and gas corporations. In most parts of the industrialised world it is a question of megabucks versus peanuts. This does not have to happen. Market barriers to energy efficiency can be overcome.

Overcoming Market Barriers

A range of interventionist measures by government, some voluntary but many taking the form of regulations of various sorts, can be used to overcome market barriers to energy efficiency.

We can regulate the way the energy industry works to ensure that the balance of investment shifts towards energy efficiency. The energy industries could market and invest in energy efficiency as well as selling energy. Consumers can pay for energy efficiency as part of their bills in the same way as they pay for energy supply.

The consumer will spend less on energy, more on energy efficiency and the energy industry can still make profits. The difference is that pollution levels and consumers' energy bills can be reduced by spending money on energy efficiency measures instead of buying energy and building pipelines and power stations.

In future the energy industry may consist of many companies who supply energy efficiency services as well as vendors of various traditional and new brews of energy supply.

There is nothing unnatural about giving priority to investments in energy efficiency; it is just that human culture has so far placed priority on supply-side investments. Energy industries have grown up in a productionist as opposed to conservationist culture.

Laws and regulations which have shaped energy industries in their present supply-oriented forms can be changed. Instead of encouraging energy companies to make money by selling energy, they can be encouraged to make money by selling energy services. Changes in cultural attitudes can result in more attention being given to energy efficiency, but given the market barriers to energy efficiency we also need policy instruments to promote energy-efficient practices and techniques.

Appropriate intervention in the market can ensure that consumers' energy costs are reduced by investment in energy efficiency. A market in energy efficiency can be created. This is being demonstrated to an increasing effect by some US utilities through the integrated resource planning system, by the Danes through their regulations which insist that fossil fuel power stations must be part of cogeneration systems and by the Dutch energy utilities which are driven by the need to meet environmental targets. I shall now examine the policies that can be used to encourage the adoption of energy efficiency techniques.

Energy Efficiency Policies

Policies to promote energy efficiency can be either 'passive' or 'active'. Passive policies will affect the energy consumer directly only if he wishes to take advantage of them. Passive policies include energy advice services, grants and tax incentives for energy efficiency.

National budgets are hard pressed and while these passive stratagems can be valuable adjuncts to other, more far-reaching policies, their effects are, on their own, likely to be fairly modest.

Active policies to promote energy efficiency will influence the actions of all energy consumers. Energy taxes are a well-known 'active' method of encouraging the efficient use of energy.

Energy Taxes

Taxes raise money for the national treasury. Energy taxes influence the 'macroeconomic' market behaviour. They can take several forms.

Taxes can be levied on primary (raw) energy supplies. A leading example of such a primary energy tax is the widely favoured but (as yet) little implemented carbon tax concept. Carbon taxes are levied on primary energy supplies according to the carbon content of the fuel. Thus a $1 per GJ tax on coal would coexist with a 57 cents per GJ tax on natural gas and a 77 cents per GJ tax on crude oil. Norway, Sweden, Finland and Switzerland have levied carbon taxes, although some of their effect has been to replace existing energy taxes. A proposal to levy an EU-wide carbon tax has been knocking around for some years, but has been repeatedly blocked by the UK and Spain in particular.

Energy taxes are frequently levied on delivered energy, for example on motor vehicle fuel or through Value Added Tax (VAT) on gas and electricity. This type of taxation has the advantage that it is difficult to evade, although it is also a generally regressive means of raising revenue. Some argue that energy taxes can be applied more progressively.[21] Many have argued that other taxes could be reduced as energy taxes and other 'resource taxes' are increased.[22]

Sometimes energy taxes are advocated as a means of raising funds for energy efficiency. In practice this is difficult to do because there is no formal link between specific taxes and specific public spending programmes. Money raised this way is going to be in competition with many other pressing claims on the public purse.

The most realistic use of energy taxes to influence energy practices is as a penalty on consumers. Taxes can pressure consumers to accept a lower level of energy services by cutting consumption, they can induce them to invest in energy-efficient equipment, or the taxes can encourage them to shift to another energy supply option which is subject to lower or zero taxation.

Energy taxes appear to be such an obvious option for encouraging energy efficiency because energy price rises in the past have led to increased spending on energy efficiency. But the oil price shocks of the 1970s were massive. Crude oil prices rose, in US dollar terms, sixfold between 1972 and 1980.

It requires very high tax increases to have major effects on energy consumption patterns. Various studies have suggested that carbon taxes would have to triple or even quadruple the price of primary energy in the industrialised world before current levels of carbon dioxide emissions were, by the end of a 15-year period, reduced by 20 per cent below present levels.[23]

The environmental impact of relatively modest tax changes on energy consumption is small. The 8 per cent VAT on domestic fuel in the UK will reduce household energy consumption by about 3 per cent below what it would otherwise have been.[24] This translates into a reduction of no more than 1 per cent of the carbon dioxide emissions produced by the whole of the economy.

As was mentioned in Chapter 5, the biggest problem with taxes is that there is a low elasticity of demand for energy with respect to price. Moreover, the effect on consumers is only partly to induce them to invest in energy efficiency. A lot of the modest reductions in demand result from consumers deciding simply to cut bills by accepting lower levels of energy services, and this will be particularly true in the case of lower income consumers.

Politically speaking, energy taxes are hard to levy in the industrial sector because of claims that it will harm competitiveness. In most countries industry can reclaim, or is exempt from, part of or all of energy taxes.

Tax increases are (especially these days) unpopular. Bill Clinton's proposal for a general energy tax called the Btu tax, made in 1993, was booted out in Congress in the face of public hostility and was replaced with a mere 4.3 cents per gallon increase in gasoline taxes.

Supporters of an energy taxation strategy often cite the high energy taxes in force in countries like Denmark and Italy. Yet such taxes were originally levied because these countries are extremely energy-dependent and have large oil import bills. In such circumstances energy taxes can help the economy by reducing the balance of payments deficit.

High Danish energy taxes levied since 1985 have helped to reinforce their cogeneration programme by encouraging consumers to switch from oil-fired heating to district heating associated with cogeneration schemes. However, the planning laws which effectively ban electricity-only fossil fuel plant have been a key element of the very rapid shift towards cogeneration in that country. The Netherlands, which has very low energy prices, is relying on regulatory methods to ensure rapid expansion of cogeneration technology.

Higher motor vehicle fuel taxes can reduce the pressure for road-building, reduce congestion on roads and make public transport more attractive as well as promoting the purchase of energy-efficient vehicles. These collateral advantages serve to diminish, although not entirely eradicate, the unpopularity of tax increases in the transport sector. The UK government has agreed to a programme of increasing petrol taxes by 5 per cent a year.

There is no doubt that high rates of energy taxes will have an impact on energy demand, but if such measures are employed solely as measures to encourage energy efficiency then energy efficiency is going to attract hostility. Politically, the demand for high energy taxes has the effect of distracting attention from measures that do not require any net increase in consumers' energy bills.

Energy taxes are an even less likely strategy in most parts of the developing world. In poor countries it is difficult enough to remove subsidies from energy supplies, never mind impose taxes.

Market-based Incentives

The most prominent use of market-based incentives for environmental purposes to date has been the US tradeable permits system for controlling SO_2 emissions. Some permits have been made available for energy efficiency improvements, although most of the cuts in sulphur emissions are likely to be achieved either through retrofitting coal-fired power stations or by using more natural gas.

The central difference between the acid emissions and the carbon dioxide emissions problem is that the former is a by-product of the fossil fuel combustion process while the latter is the main product. This makes it difficult to apply a tradeable emission permits scheme to the carbon dioxide problem. Nevertheless, the US Environmental Defense Fund has proposed a tradeable efficiency credits scheme as a flexible alternative to energy efficiency standards,[25] but as yet no such scheme has been implemented.

There have been some proposals for a tradeable carbon dioxide emissions permits system operating on an international level. For example, Mike Grubb has suggested that countries could be issued with permits according to their population sizes and the number of permits available could be steadily reduced.[26] Currently there is no international consensus for such a scheme although there have been suggestions in the US that US companies could finance cuts in emissions in developing countries in exchange for increasing domestic emissions. A lot of environmentalist groups oppose any notion that people ought to be able to 'buy' the right to pollute. Regulation is the better-trodden route to ecological salvation.

Regulation

Regulations take decisions on behalf of the consumers, for example, over maximum permissible emissions of pollutants from their motor cars. These days regulations are unfashionable. They are often referred to by economists as 'command and control' mechanisms, but they are often unavoidable.

Regulations are said to prevent the market from allocating resources efficiently. Of course this assumes that markets are (more or less) perfectly competitive, which they rarely are. If regulatory means can be used to remove market barriers to investment in energy efficiency and if they can create a market in energy efficiency then regulatory tactics could make energy markets more, not less, competitive.

Perhaps the simplest regulatory means of enforcing energy-efficient practices is to set minimum standards for energy efficiency in buildings and energy-using equipment.

Energy Efficiency Standards

Energy efficiency standards have been part of building regulations in industrialised parts of the world for several decades. They have been gradually tightened up, and can be tightened much farther. Such standards have tended to be stricter in colder countries (like Canada and Scandinavian countries). Yet new houses built to standard in the UK consume over a quarter more energy than houses built to standard in The Netherlands despite the similarity of climates and the fact that Dutch natural gas prices are similar to those in the UK.

Energy efficiency standards can also be set for all standardised pieces of energy-using equipment. Some countries set efficiency standards in the 1970s and early 1980s as a response to the oil crises, but then as oil prices fell interest slackened off and it is only recently that interest has revived as a result of pressure from environmentalists.

The US began setting standards in 1976, although these have been greatly strengthened under the 1992 Energy Policy Act. Standards have been set for most common energy-using appliances ranging from televisions to air-conditioning, and from water heaters to furnaces. Some states set stricter standards than those set under federal law.

Studies made under the auspices of the US government indicate that the new standards would, overall, produce net cost benefits for consumers.[27] The Electric Power Research Institute (EPRI) estimates that the impact of the efficiency standards and market efficiency improvements will reduce US growth in electricity demand by 0.5 per cent per year in the period 1990 to 2000.[28]

Energy efficiency standards for electrical and other appliances have also been set in some developing countries, notably in Taiwan and Brazil. The Taiwanese government estimates that electricity appliance efficiency improvements between 1981 and 1988 have reduced electricity generating capacity requirements by around 5 per cent between 1981 and 1988.[29]

Despite a lot of talk in the European Union about setting European-wide efficiency standards (under the SAVE programme) little action has actually emerged apart from an energy labelling scheme for fridges and freezers and a standard on boiler efficiency from which the UK opted out. Indeed, the net effect of the EU has in some respects been to make things worse. In 1992 the EC Commission ordered the Dutch to halt plans to set standards for fridges saying that it was contrary to the free competition rules of the EU. The Commission agreed to bring in EU-wide proposals, but meaningful action has failed to materialise.[30]

Some people argue that agreements with industry are better than regulations. In reality the difference between regulation and agreements can be very thin. The West German government reached agreement with fridge manufacturers to improve the energy efficiency of their products by 20 per cent between 1980 and 1984. The Japanese reached a similar agreement about fridges and air-conditioners which was backed up by an order made under their 1979 Energy Conservation Law. Was this a regulation or an agreement? No significant agreements or regulations about energy efficiency have been forthcoming in the UK.

The main advantage of efficiency standards is that they can eliminate the most energy-inefficient models, although resistance from manufacturers acts to prevent more radical action.

If you want something done to correct the imbalance in investment between energy efficiency and energy supply then energy utilities are well placed to act. Since the energy crises of the 1970s there have been growing pressures to make US utilities plan properly to deliver services to the consumer according to the least-cost mix of resources. This has evolved into a system called Integrated Resource Planning or IRP.

Integrated Resource Planning

IRP (formerly known as 'least-cost' planning) tries to determine the lowest-cost means of supplying energy services to the consumer using the most cost-effective mix of not only existing (mostly conventional) fossil fuel and nuclear plant but also cogeneration, energy efficiency and renewable energy. IRP is best known to energy efficiency enthusiasts for its association with demand side management, or DSM.

DSM has been defined in many ways, but I use it here to mean techniques involving the marketing of and investment in end-use energy efficiency techniques. DSM is a set of techniques that is independent of the IRP system itself.

The Public Utility Commissions (PUCs) are the arbiters of the IRP process which involves submissions from energy suppliers, consumer (ratepayers) groups and environmental groups to regular reviews of policy. The strongest DSM programmes have been implemented in the electricity industry. DSM activity in the natural gas sector has been much weaker.

The principal aim of DSM programmes has tended to be the reduction in peak demand (which is expensive to supply) rather than energy savings per se. Nevertheless, large energy savings are claimed by utilities and the emphasis seems to be shifting towards energy saving as the prime aim. New England states, New Jersey, New York and California project that around 30 to 35 per cent of growth in energy demand will be met by DSM in the 1990 to 2001 period. These states spend around 2 per cent of their annual electricity receipts on DSM-related activities.[31]

However, many states have only weak or even non-existent IRP systems. In West Virginia, for example, the main utility has a 50 per cent over-supply capacity, a problem which is evident in many states which had over-profligate power station expansion activities in the 1970s and 1980s. In some cases the 'avoidable costs' (the costs we would avoid paying by not buying the energy) of extra electricity might be only 1–2 c/kWh, but even at these prices there is room for cheap DSM measures which are frequently not taken up. After the year 2000 power station retirements will increase. Avoidable costs will increase considerably with the demand for replacement capacity and the market for 'least-cost' DSM will increase in a corresponding fashion.

DSM programmes have been championed by environmentalists (both to cut pollution and constrain the need to build new power stations and transmission lines), but they have been implemented by utilities for the purposes of reducing consumers' bills.

There are several methods used (in different states) to ensure that energy sales are 'decoupled' from income to utilities so that the utilities can make money out of DSM as well as selling extra units. One method is to assess what electricity demand is likely to be over the next few years and set rates sufficient to give the utility a reasonable return; another is to allow a good return on energy efficiency investments to compensate for sales losses; and a third is to allow the utilities to earn revenue according to their numbers of customers. 'Decoupling' can work because the utilities are monopoly retailers of energy.

Unit prices may rise by a small proportion[32] but the volume of energy used by the average consumer falls by a greater amount, producing a decline in the average consumers' bill.

The main instrument of DSM programmes is rebates (subsidies) to energy consumers to induce them to adopt energy-saving techniques or equipment. Rebates are given to people who buy showerheads that use less water (and thus less energy to heat the water), energy-efficient air-conditioning equipment or refrigerators, variable speed drives in industry, energy-efficient light fittings, energy audits and many other things. The rebates act as both marketing tools and methods by which energy efficiency measures are made cheaper. The biggest area of activity so far has been in the commercial sector.

EPRI says that the annual rate of increase in US electricity demand between the years 1990 and 2000 is being reduced from 2.3 per cent to 1.4 per cent by a combination of DSM and improving energy efficiency standards. This combination will meet 10 per cent of total electricity demand by 2000.

EPRI reports that DSM is very cheap. It costs around 2.9 cents/kWh (1.9 p/kWh) compared to the cost of power from CCGTs of 4.3 c/kWh (7 per cent discount rate, including an allowance of 0.2 c/kWh for losses in transmission that would be avoided by DSM).[33]

If DSM in all states was brought up to the levels enjoyed by the top few, and if energy efficiency standards continue to be upgraded, then growth in electricity demand could be further reduced. Cogeneration, when substituted for coal-fired plant, will reduce emissions (especially if the overall energy efficiency of the schemes is high). Continued refinement of DSM techniques and more spending on DSM will advance the process further, leading to absolute reductions in carbon dioxide emissions.

In the most advanced DSM states, such as those in the New England region, new, competitive, low-cost means of delivering DSM are being developed. Groups like the Boston-based Conservation Law Foundation (CLF) are working with utilities and product manufacturers on a range of techniques. For example, they are asking manufacturers of drinks vending machines to improve the energy efficiency of their products. Programmes involving the funding of research into and the installation of technologies such as energy-efficient commercial air-conditioning equipment are also being implemented. The CLF believes that this 'surgical' approach involving concentration on specific technologies and the regular upgrading of energy efficiency standards is the best way forward. Improved standards and DSM measures can interact to accelerate efficiency improvements.

Many independent energy service companies prefer so-called 'demand side bidding' systems. These involve the issuing of contracts to supply kWh of energy savings. The companies bidding the lowest amounts are awarded the contracts and they are paid rates for kWh of energy savings analogous to payments for kWh of energy. However, early attempts to implement these schemes have been criticised on the grounds that the programmes deliver fewer energy savings than promised.

With the opportunities for computerised control of energy used in industrial processes expanding, the cost-effective opportunities for DSM are likely to grow, especially as more consumers become aware of its energy-saving potential.

A lot of the advance in cogeneration is owed to PURPA, but in states like California the IRP process has set aside blocks of cogeneration contracts which are filled by rounds of competitive bidding. This has, as we have seen earlier in the chapter, evinced massive quantities of cogeneration at very low prices. These are prices with which conventional new capacity, offered by the utilities, cannot compete. In practice many NUGs are heavily financed by utilities themselves.

Under IRP procedures, DSM and cogeneration schemes are implemented if they are cheaper than the avoidable costs of energy supply. They are therefore, by definition, cheaper than conventional means of energy supply, a fact which completely contradicts the assertions usually made about environmentally sensitive options being expensive.

Despite the progress that has been made, IRP programmes are seriously threatened with destruction by deregulation, or 'retail wheeling' as it is known in the US. Retail wheeling means that utilities will lose their monopoly status and the power lines will be opened up to enable other suppliers to retail electricity directly to the consumer. Utilities will now have to compete on unit prices. There will be strong pressure to cut DSM investment programmes to allow them to do this.

The rationale for deregulation is that it increases the efficient use of resources though competition. Yet it is only supply-side competition and does nothing, on its own, to overcome the market barriers to the provision of DSM. Unless special action is taken to sustain a market in energy efficiency, cost-effective energy efficiency measures will not be taken up. Energy suppliers, who seek to make returns on their investments in energy supply capacity, are not going to start competing with each other to persuade and subsidise consumers to use energy-efficient equipment.

A way of maintaining the flow of funds for energy efficiency in the context of deregulation is to charge all electricity suppliers using the transmission and distribution network a levy per kWh of energy supplied. The money that is derived from the levy can be used to fund DSM activities. Institutions could then be established to invest this money in energy efficiency. I shall comment further about how this can be done when I examine how the UK's own fledgeling energy efficiency levy system could be used more effectively.

Environmental Targets

The Netherlands is, with the exception of the Danish cogeneration programme, about the only European country at present with a coherent method of regulating its utilities to achieve energy efficiency objectives. The municipally controlled Dutch electricity and gas distribution companies are on course to achieve targets of reducing 1990 levels of carbon dioxide emissions by 3 to 5 per cent by the year 2000 (see Figure 7.4). This amounts to a reduction of 20 per cent of the emissions that would have been produced in the year 2000 in the absence of the programme. The target has been set by the Dutch government. A limitation to the programme is that this leaves out a great deal of the energy economy, including the transport sector and much of the industrial sector.

The Dutch are keener environmentalists than most, but there is considerable self-interest involved in their energy efficiency policies. They have some prized gas reserves around Groningen and they want to be able to rely on this resource for a long time to come.

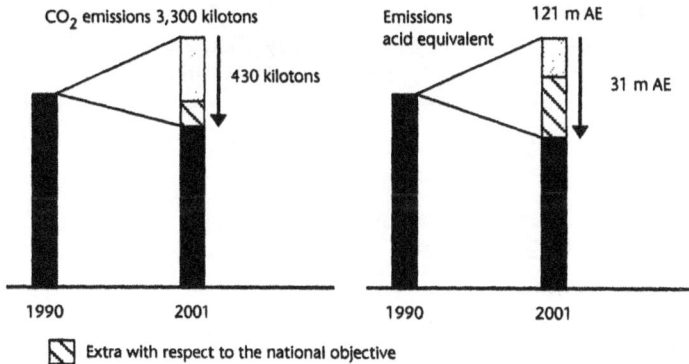

Figure 7.4 Projected Emissions Reductions for Amsterdam Energy Company

Note: acid equivalent is a measure of the increase of the acid rain resulting from emissions of NOx or SO_2: one AE stands for 46 gram NOx or 32 gram SO_2. The target has been revised upwards since this chart was originally published

Source: Amsterdam Energy Company, *Energy Planning in Amsterdam*, 1992, p. 5

The energy companies are allowed to increase unit prices by 2 per cent in order to finance some of the energy efficiency programmes organised by the energy companies. Energy consumption will be reduced and, according to the energy companies, total bills for the average consumer will decline.[34]

A key difference between the programmes being organised by Dutch and US utilities is that the Dutch energy companies are investing in overall energy efficiency natural gas cogeneration. The cogeneration programme contributes around 40 per cent of the emission reductions required under the utility targets.[35]

Cogeneration projects are supported by capital grants (17 per cent in 1993) from the government, although even without the grants the measures would still be very cheap. The level of the grant is being reduced.

The techniques used to market the end-use efficiency measures are broadly similar to US-style DSM. For example, about $500 (£330) is given as a subsidy for every kW of electricity capacity saved by installing efficient light fittings. These are reflectors that enable the lighting system to distribute more light with fewer fluorescent strips.

In Amsterdam around £18 is given for a purchase of one of the most 20 per cent energy-efficient models of fridges, provided that the consumer brings in the old one. The companies buy new boilers for consumers, and reclaim the investment by charging rent, and so on.

The Dutch system relies on energy distributors having a monopoly. It is very simple in that it relates directly to environmental objectives. This means, for example, that the companies have a clear obligation to install high overall energy efficiency cogeneration.

Rules Favouring Cogeneration

The largest number of cogeneration programmes in the world has probably been set up under the US PURPA legislation, the shortcomings of which I have already described. Potential small cogeneration schemes may achieve high levels of overall energy efficiency, but their backers may have limited access to capital and may not be able to negotiate as favourable fuel supply contracts compared with larger but more inefficient power projects.

This problem could be overcome if 'tranches' (batches) of electricity supply contracts could be set aside for cogeneration schemes with high levels of overall energy efficiency. Non-utility generators would thus have an incentive to group relatively smaller projects together so that they could enjoy better access to capital and negotiate lower fuel prices. However, it seems that the practice of holding auctions for tranches of cogeneration contracts is being ended as part of the move towards retail wheeling in states like California.

In Denmark planning laws approved in 1979 have made it virtually obligatory for people to have their heating supplied through district heating networks. Since 1990 it has been mandatory that new electricity production should supply heat to the district heating networks. This has, in effect, banned electricity-only fossil fuel power stations. Capital grants are given to people who set up cogeneration schemes.

In the UK environmentalists are calling for planning laws to be altered to give priority to cogeneration schemes. The Combined Heat and Power Association wants cogenerators to be paid fair rates for the power they sell to the grid. Cogenerators receive only the relatively low 'pool' price for their electricity while the main power station operators are paid much higher rates as part of premium price contracts agreed with the regional electricity distribution companies (RECs).

Energy Efficiency Levy

In the UK a system is evolving whereby a small percentage of energy bills (a levy) is used as the central means of paying for the energy efficiency investment programme in the electricity and natural gas sectors. To date some 'pilot' schemes have been funded in the gas sector. The main initial measure has been the funding of investments to the tune of £25 million a year in the domestic sector of the English and Welsh electricity supply market from 1994 to 1998. This is small compared to the Dutch programme. For example, the Amsterdam Energy Company alone is spending £22 million a year on a population of just 700,000.

The Energy Savings Trust (EST) has been established with the task of managing the UK programme. The measures are being implemented by the regional electricity distribution companies (RECs). In 1993 the EST published plans for extending the programme to total £400 million of annual investment by the year 2000. These levels of investment would produce energy savings that are cheaper than the alternative measures of energy supply, but institutional pressures have delayed the implementation of this programme.

The EST's programme has been designed to fit in with the way the electricity and gas sectors are being regulated (and, progressively, deregulated). Consequently there are problems. Whereas energy efficiency investment is, in the case of the US and Dutch programmes, financed by utility borrowing, the UK's programme is being funded directly from the funds raised by the levy. This has the result that the average consumer has to wait four years before bills decline, whereas the average bill would decline from day one if the investments were financed by borrowing. In the latter case the money gained from energy savings would more than cover the loan and interest repayments needed to service the debts.

The scheme has been organised in a fashion that reduces short-term benefits to the consumer because, under existing regulatory arrangements, RECs have no financial interest in borrowing to finance the programme. For a start the RECs earn more money by selling more units so that the energy suppliers actually lose money by implementing the programme! The move towards deregulation in electricity and gas is making things worse.

The deregulation process discourages the energy utilities from borrowing money to finance energy efficiency since they cannot be certain that the customer base (from whom the utilities would have to reclaim the borrowings) will still be around. They could be merrily buying their energy from someone else who does not have to pay back money borrowed to finance energy efficiency.

The EST's scheme has been attacked as being 'regressive' by the the government-appointed gas regulator, OFGAS, despite the fact that it is the regulations set by OFGAS (and the electricity regulator OFFER) that have forced the EST to organise the scheme in an imperfect manner.

The levy could be used more effectively as an assured stream of income that could be used to repay money borrowed by the EST to finance investments in energy efficiency. There are various ways of organising this.

The EST could give funds to manufacturers to make energy-efficient appliances available at low prices, as has been done with compact fluorescent lamps (with electronic ballasts) in California and also for a limited period in the UK.

Giving the incentives to the manufacturers (under conditions where manufacturers received funds acccording to the numbers they sell) reduces the final retail price much more than giving the incentive to the consumer. This is because the value of the rebate is effectively increased by the percentage 'mark-up' between the manufacturing and retail stages.[36]

Another way of using the levy would be to afford a guaranteed stream of income to a parallel set of 'Regional Energy Efficiency Companies' (REECs). The REECs could operate either as public bodies that could borrow on the basis of a government guarantee or as private sector organisations funded by commercial banks and shareholders. The REECs would organise the investments in energy efficiency.

The energy efficiency levy can thus be used to strengthen the market for energy efficiency services. These services can then compete on the same terms as those enjoyed by energy suppliers.

Some have suggested that energy efficiency should be funded out of taxes. Apart from the fact that this would severely limit the availability of funds for energy efficiency, the point is that investment in energy efficiency is just as much a justifiable way of meeting consumer needs for energy services as investment in power stations, gas pipelines or the purchase of raw energy supplies. Thus energy efficiency (like energy supplies) can be justifiably funded out of revenues collected from sales of energy services.

Among the excuses for the lack of an energy efficiency programme in the UK is the idea that the structure of the utilities is different from that obtained in the US and the notion that electricity prices are higher in the US. If anything the UK structure (where there is a clear separation between distributors and generators) makes the distributors less, not more, reliant on the generators for energy. The RECs can substitute DSM for energy, just as is done by the Dutch who avoid buying power from their central generators. The costs of power from new power stations are likely to be much the same in the UK, the US and The Netherlands, given that energy prices are broadly the same in all three countries.[37]

A thoroughgoing energy efficiency programme in the UK will be based on US- and Dutch-style DSM methods.

Criticisms of DSM

DSM programmes have been criticised because 'non-participating' customers who do not receive subsidies for energy efficiency must share the costs of making these subsidies. In addition, some of the people who are given subsidies to buy energy-efficient products are people who would have bought them anyway (so-called free riders).

DSM advocates say the increase in bills suffered by non-participators is small and more than compensated for by reductions in bills to the 'average' consumer. Moreover, 'free riders' may be offset by so-called 'free drivers'. Free drivers benefit from the greater priority given by manufacturers to designing energy efficiency products even in the case of appliances that do not achieve high enough performance to qualify for subsidies.[38]

An alleged example of the 'free driver' phenomenon was given to me by a Dutch utility executive. At the beginning of the Amsterdam Energy Company's lighting programme a subsidy was given only to those consumers who bought the 20 per cent most efficient light fittings, but within three years half the fittings on the market were as efficient as the original 20 per cent although the subsidies were still being given only in respect of the (new) top 20 per cent. This means that many people who are not being subsidised are nevertheless buying products that are more efficient because of the market forcing effect of the subsidy programme.[39]

While there would be little point in schemes that consisted totally of free riders, the free rider problem as well as the problem of cross subsidisation by non-participating customers are phenomena that have always existed on the energy supply side. Why should someone who is using energy efficiently subsidise people who are using their energy inefficiently through paying higher prices to pay for extra and more expensive sources of supply?[40]

There are various ways in which energy efficiency programmes could compensate non-participating customers, for example through schemes where those achieving reduced bills through subsidised energy efficiency investments shared their savings with other consumers.[41]

However, given the gross inequities and inefficiencies of the current system wherein considerable market barriers to energy efficiency exist, it seems reasonable to argue that the emphasis should be on establishing a comprehensive system of funding of energy efficiency and then fine-tuning the system.

The fact that DSM programmes benefit some consumers more than others is also matched by the fact that some energy supply developments will benefit some people more than others. Why should one area benefit from the increased jobs resulting from new power plants and not others? The issue should be whether the community as a whole gains from power stations or energy efficiency programmes.

There are some criticisms of DSM programmes that may be justified. In many US states the utilities are encouraged to invest in DSM by being allowed (by the regulators) to earn a high rate of return from DSM investments. This type of incentive does not encourage the utilities to maximise the efficiency of DSM investments since they can still earn more money from selling more units.

Indeed, many energy service companies who are hired by the utilities to do energy audits feel that the money could be more effectively spent. Because of such criticisms DSM advocates have, in recent years, tended to recommend a different form of regulation. This can involve decoupling income from sales by allowing the utilities to earn revenue according to the number of customers, regardless of the energy sales.

Nevertheless, despite the inadequacies of DSM that do exist, DSM critics have been unable to question the low costs of DSM. The alternative strategies suggested by DSM sceptics, such as agreements with manufacturers to make energy-efficient appliances, are strategies that are valuable additions to, rather than substitutes for, DSM programmes.

Probably the biggest criticism of DSM is simply that it has not been deployed widely enough. The Netherlands apart, utilities in Europe have been very slow to take up DSM.

The potential impact of DSM and high overall energy efficiency cogeneration could be very large indeed. The Dutch system (and, in the case of cogeneration, the Danish system) demonstrates the possibilities for achieving significant reductions in carbon dioxide emissions at negative cost to the average consumer.

Improved methods of delivering end-use energy efficiency, the use of levies to finance energy efficiency programmes, the adoption of rules favouring high overall energy efficiency cogeneration, and a general reduction in the costs of energy efficiency techniques, could all enhance these possibilities. An environmentalist-led strategy might set the size of an energy efficiency 'levy' according to the desired level of emission reductions. The degree that this sort of tactic needs to be employed depends very much on estimates of how demand for energy services are likely to change.

Energy Demand in the Future

It is quite legitimate to argue that projections of future energy demand are exercises in fantasy. The trouble is that people have to try to make them in order to examine the impacts of different energy policies. Governments tend to give much higher estimates than environmentalist groups.

For example, in 1990 the UK government estimated that carbon dioxide emissions would, if no action was taken (in a 'business as usual' scenario), be some 30 per cent higher in the year 2005 than at the end of the 1980s. Others disputed this figure as political gamesmanship deployed to make the official 'stabilisation' of 1990 emissions by 2005 target look radical. In 1994 the Cambridge Econometrics Group projected that without any new measures (beyond the expected new CCGT power plants and the tax changes that had already been announced) carbon dioxide emissions in the year 2005 would actually be 6 per cent lower than in 1990.[42]

The world as a whole actually saw a drop in energy use between 1990 and 1992. Recession in the West and collapse in Eastern Europe was chiefly responsible for this turn of events. However, the recent underlying trend has been for world energy use to grow much more slowly than would have been anticipated in the 1960s. World energy use expanded by only 10 per cent in the 1986 to 1992 period compared to a 38 per cent increase in the 1967 to 1973 period.

The frequent reaction from analysts schooled in the belief that the 1945–73 period represents normality is to dismiss recent trends as the result of a set of atypical political occurrences. Yet closer inspection reveals that it is the 1945–73 period which had much higher rates of world economic and energy growth than either before the start of the period or after its end. Indeed, the rate of economic growth in the leading industrial states during the 1950 to 1973 period was over twice as high as the average rate of economic growth since 1720.[43] These trends can be seen in Figure 7.5.

Are we seeing a reversion to type, with the more moderate and intermittent rates of economic growth observed since 1973?

Certainly the growing nationalisms of the world, combined with heightened clashes based on religion, do not suggest an especially stable period ahead. Projected rates of consistent high economic growth in the already industrialised world should be treated with scepticism. The growth in geopolitical instability may not only produce crises that depress the rate of economic and energy growth, but it is also likely to bring energy security issues more to the fore.

An important driver of energy growth is the rate of increase in world population. This is part of the explanation for the big increases in energy consumption experienced in the developing world compared to, say, OECD countries. (This can be seen in Table 7.7.)

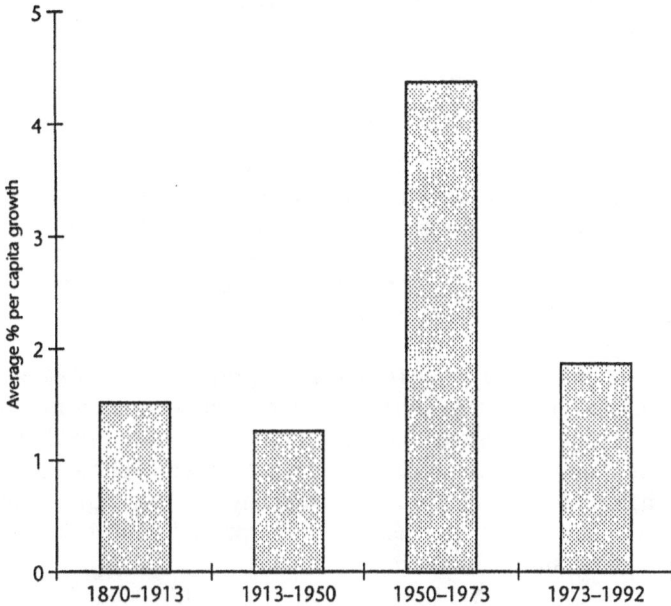

Figure 7.5 Average Annual Economic Growth since 1870 in Selected States
(US, Japan, Germany, France, UK, Russia/USSR)

Source: *Financial Times*, 9 March 1994, p. 14

In countries like the US each percentage point of (GDP) economic growth has been achieved, in recent years, with much less growth in energy consumption than occurred before the oil crises. According to US Department of Energy statistics, the 1963 to 1973 period saw average annual US GDP growth of 4 per cent while the annual increase in US energy consumption was 4.4 per cent. In the 1983 to 1993 period there was annual US GDP growth of 2.8 per cent compared to an annual increase in US energy consumption of just 1.8 per cent.

The Brundtland Report hoped that increasing per capita living standards would reduce the high population growth levels existing in the developing world, thus mirroring the decline in population growth observed in the industrialised world. This has not yet happened. In many poorer nations much or all of the economic growth has been cancelled out (in per capita terms) by population increases.

Table 7.7 Changes in World Energy Consumption, 1982–92 (MTOE)

	1982	1992	% change
OECD	3,451	4,128	+20
E. Europe and former Soviet Union	1,531	1,513	−1
Lesser developed countries	1,386	2,154	+55
World	6,368	7,795	+22

Source: *BP Statistical Review of World Energy*

On the other hand, there is a growing appreciation of the need to check population growth in order to achieve significant improvements in living standards. There is a definite slowdown of population growth in countries such as Kenya and Bangladesh. This change has been widely attributed to improved women's rights including access to better healthcare, improved education and greater social status for women.[44]

Another analysis says that population growth levels off when there is no more productive land (and associated fuelwood resources) to be 'captured' by the peasants. When most resources become privately owned the incentive to have many children in order to capture the natural resources is removed.[45] A corollary to this argument is that if natural habitats like the rainforests can be protected then there will be less room for new settlements and population growth. Individual living standards could improve since the fruits of economic development would be shared among fewer people.

If population growth rates do decline towards those exhibited in the already developed world then much of the pressure for increased energy consumption in the developing world would be taken away. Such a trend towards declining energy growth rates could be reinforced by other forces pushing the world along a low energy path.

The oil crises may have shifted energy consumption on to a new, less energy-intensive path by focusing attention on alternative materials and methods that use less energy. These new techniques are still being employed despite the fall in oil prices. Besides these trends, there is a general shift in growth patterns away from traditional heavy, metal-based industries towards much less energy-intensive industries and services based on 'value added' services and information technology. This change improves the efficiency with which energy is used and is part of a trend called dematerialisation.

Dematerialisation

Bernardini and Galli have defined dematerialisation as having two basic postulates:

> (1) that the intensity of use of a given material (or energy) follows the same pattern for all economies, at first increasing with per capita GDP, eventually declining.
> (2) that the maximum intensity of use declines the later in time it is attained by a given economy.[46]

This trend is illustrated in Figure 7.6 below. Energy intensity means the amount of energy used to produce a unit of GDP. Energy intensity tends to decline once a nation has built its basic industrial infrastructure. Developing nations will complete this process with much lower levels of energy intensity than were achieved by the West when its infrastructures were established. Moreover, the nature of industrial production itself has changed and continues to change in a low-energy direction.

We have deeply engrained beliefs about the inevitability of expansion in energy use. But has the industrial revolution with its emphasis on energy-intensive smokestack industries producing hunks of metal largely come to an end? Are we now seeing a move into an age where the emphasis is more on designing products for individual needs than on producing more and more mass, standardised products? Quality instead of quantity? Much of the economic development may in future be in terms of 'value added' better design, products that have the potential to perform many functions, and a range of new services. Such 'quality'-driven growth is inherently less energy-intensive than growth in the number of traditional industrial products.

There are several reasons for this. First is the development of new types of material (such as carbon fibre) which allow production lines to become more flexible and products less bulky. Second is the development of information technology which tremendously increases the degree to which resources can be efficiently utilised and greatly expands the number of services that can be provided for an affordable price. Third are cultural changes. These result partly from the growing affluence in the West and the saturation of demand for basic products and services. These cultural changes

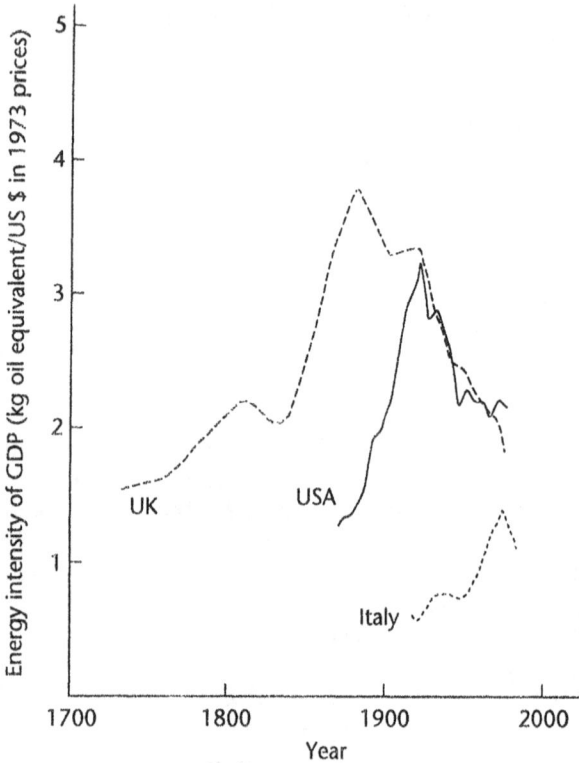

Figure 7.6 Long-term Trends in the Energy Intensity of GDP
in the UK, the US and Italy

Source: O. Bernardini and R. Galli, 'Dematerialization. Long-
term Trends in the Intensity of Use of Materials and Energy',
Futures, May 1993, p. 435

favour products which match individual tastes and also production
that involves low levels of pollution.

In the past the development of technologies such as the steam
engine, the electric motor and the internal combustion engine
created enormous scope for increased economic growth. These
are energy-intensive technologies. However, information
technology is conspicuous, as William Walker puts it, for its
extremely 'abstemious' use of energy.[47] Moreover, as has already
been described, information technology presents us with enormous
potential for improving the efficiency with which energy is used.

The increasing concentration of investment in information technology means that the new wave of economic development could prove to involve much lower energy intensities than previous eras of economic development. The dramatic recent increase in investment in information technology in the US is shown in Figure 7.7.

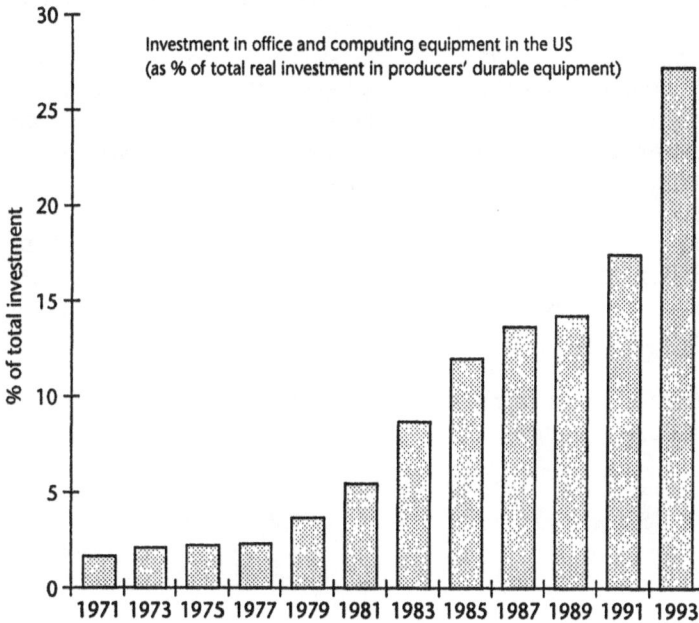

Figure 7.7 US Investment in Information Technology

Source: *Financial Times,*14 March 1994, p. 20

This 'low energy intensity', information-technology-led growth is likely to reinforce other trends constraining the rate of energy consumption.

I have already described the large impact that interventionist policies favouring a low-cost delivery of energy efficiency services could achieve. Pressure from environmentalists is likely to intensify the cultural shift towards giving greater prominence to satisfying our needs by conserving rather than by using up finite resources. The need to meet specific environmental targets will focus attention on the cheapest means of attaining these objectives.

It is likely that the adoption of interventionist policies aimed at achieving energy efficiency will become more and more comprehensive, perhaps first in some Western countries, but eventually throughout the world.

If 'low energy' industrial trends and a slowing down of world population growth patterns reinforce cultural shifts and regulatory policies that favour energy efficiency, it may not be fanciful to talk about the possibility of world stabilisation of carbon dioxide emissions within the next 30 years.

This involves considerable reductions in emissions from already industrialised countries and a slowdown in the growth of emissions from the rest of the world. This can be acheived in the context of no extra costs to the energy consumer, low energy prices and sustained and sustainable economic development.

Beyond this it may be possible to effect radical reductions in global fossil fuel energy use only by the development and use of non-fossil fuels. I shall examine the possibilities for this in later chapters. However, we have not yet finished with considering a major contributor to the future shape of energy growth and the future level of emissions. This contributor is transport, and in particular the motor vehicle. Can the levels of air pollution it produces be reduced?

8

Reducing Pollution from Motor Vehicles

This chapter deals with the energy-related pollution that results from motor vehicle use and the practical methods by which this pollution can be abated.

There are many means of doing this, although all of them have to deal with people's demand for cheap, quick, convenient and flexible forms of travel. This, for expanding numbers of relatively affluent humans, often means the car. Over the last 20 years the number of passenger kilometres travelled by car has roughly doubled in most western industrialised states (see Figure 8.1). The exception is the slower increase in the US which already has much higher levels of car travel per person than other nations.

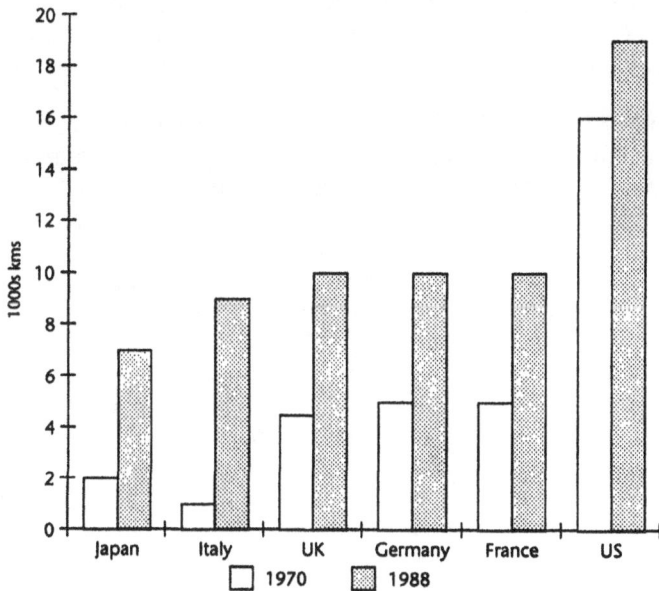

Figure 8.1 Car Travel, 1970 and 1988, per capita for Selected Countries
Source: IEA/OECD, *Cars and Climate Change* (Paris, 1993), p. 36

The main emissions produced by motor vehicles are, in order of bulk, carbon dioxide, carbon monoxide, nitrogen oxides, hydrocarbons (HC, often described as volatile organic compounds or VOCs), particulates and sulphur dioxide. VOCs interact with nitric oxide to produce photochemical smog, of which ozone is a major component.

Particulates can be carcinogenic and may, as discussed in Chapter 2, be a major cause of many respiratory and circulatory ailments. Volatile organic compounds (VOCs) include polyaromatic hydrocarbons (PAH), such as benzene, which are carcinogens. Benzene is produced from car exhausts and it is also emitted while fuel tanks are being filled. Low-level ozone produces respiratory problems, although it also contributes to some extent to global warming.

NOx and SO_2 emissions from motor vehicles cause acid rain which results in death to fish, trees and possibly harms humans.

Roadside concentrations of carbon monoxide can cause headaches and chronic exposure may exacerbate heart disease.

Carbon dioxide is chemically inert although it makes a large contribution to new global warming.

Methane, although a hydrocarbon, is not counted as a volatile organic compound, and it is relatively inert. However, it contributes to global warming at a much higher effect per molecule than carbon dioxide.

Tables 8.1 and 8.2 show how the quantities of the various types of emission have changed since 1970. Significant reductions in some emissions have been achieved in the US, whereas emissions have increased in the case of the UK (which is broadly typical of Western Europe). On the other hand, as can be seen in Table 8.3, the sheer amount of car use in the US means they have higher per capita emissions than the UK. Table 8.3 shows that Japanese emissions are much lower than in either the UK or the US. The Japanese certainly use fuel-efficient vehicles with exhaust controls, but they also drive less than the British, despite a much higher per capita GDP. It is important to note that the total amount of emissions is a combination of car use and emissions per car.

A number of factors explain the differences between these three countries. Most of the US has developed in the era of the car and an affluent car-dependent culture has emerged. This is reflected in the sprawling, low-density cities and in a political culture that keeps fuel prices low. By way of contrast, the UK and Japan are much more densely populated countries (see Figure 8.2) with an urban structure that largely predates mass car use.

Table 8.1 Emissions from Mobile Sources in the US, 1970–91

	1,000s tonnes					m tonnes
	SO_2	NOx	CO	VOC	Part.	CO_2
1970	610	8,450	96,850	12,760	1,180	1,080
1980	900	12,460	77,380	8,100	1,310	1,310
1991	990	7,260	43,490	5,080	1,570	1,489
% change 1970–91	+62	–14	–55	–60	+33	+38

Note: CO is carbon monoxide; CO_2 is carbon dioxide; part. is particulates
Source: *OECD Compendium of Environmental Data* (Paris: OECD, 1993)

Table 8.2 Emissions from Mobile Sources in the UK, 1970–91

	1,000s tonnes					m tonnes
	SO_2	NOx	CO	VOC	Part.	CO_2
1970	200	828	2,975	698	108	99
1980	117	975	4,145	900	123	107
1991	123	1,578	6,057	997	212	142
% change 1970–91	–39	+91	+200	+43	+96	+43

Source: *OECD Compendium of Environmental Data* (Paris: OECD, 1993)

Table 8.3 Emissions from Mobile Sources per Head of Population, 1990 (kg per head)

	SO_2	NOx	CO	VOC	Part.	CO_2
US	4	30	172	20	6	6,000
UK	2	30	100	17	4	2,500
Japan	2	4	NFA	NFA	NFA	2,000

Note: NFA Means 'no figures available'
Sources: *OECD Compendium of Environmental Data* (Paris: OECD, 1993);
Economist Book of Vital World Statistics (London: Economist Publications, 1990)

Nevertheless, the trend in recent years in countries like the UK is for the average length of car journeys, rather than their number, to increase.[1] Improvements in the road system have encouraged a tendency towards 'suburban sprawl' and a general lowering of urban densities.

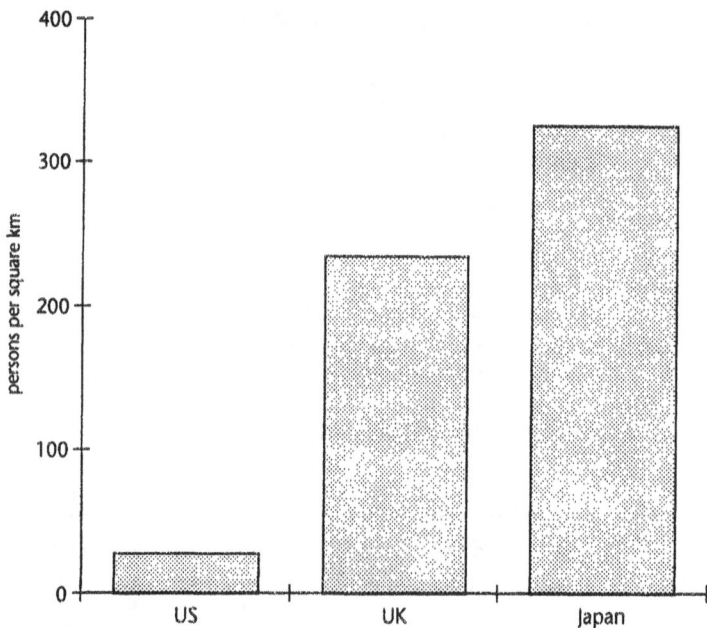

Figure 8.2 Population Densities (persons per sq. km)

Source: European Commission, *Map of Europe* (Brussels: EC, 1988)

Emission Reduction Targets

The first set of targets is concerned with a reduction in the local and regional pollution caused by transport. In areas which are subject to temperature inversions and where there are large concentrations of motor vehicles, these targets will need to be stricter than in other places if all zones are to meet World Health Organisation standards.

The second target is a reduction in carbon dioxide emissions. Given the difficulty, indeed, impossibility, of removing carbon dioxide from motor vehicles by 'end of pipe' methods, this is a particularly big problem. As has been observed earlier, if global carbon dioxide emissions are at least to start to decline in the foreseeable future, there would need to be considerable reductions in carbon dioxide emissions from transport in the developed world to offset the increases in newly industrialising nations. Even if a target of 'stabilising' carbon dioxide emissions from transport is accepted (and this is far from being the case in most countries), this puts extra pressure on other sectors to reduce their emissions. Scientists want global reductions in carbon dioxide emissions of 60 per cent or more compared to current levels.

This target appears to conflict with official projections of increases in motor car use. In 1989 the UK government projected a growth in traffic of 83 to 142 per cent between 1989 and 2025. Even in the highly motorised US traffic growth is anticipated. The US DOE projects that the number of miles travelled by light duty vehicles will increase by 41 per cent between 1990 and 2010.

A third target is a reduction in reliance on imported oil, although the amount of priority given to this target will vary according to the level of oil dependence and the price of oil. The US is becoming increasingly oil-dependent and, as was discussed in Chapter 3, oil prices may be volatile in the future.

Emission reduction targets should not be considered in isolation from the other pressures. Some are environmental; for example there are campaigns against building roads which destroy homes and countryside. On the other hand, governments are under pressure to reduce congestion on the roads.

There are various strategies that can be employed to achieve these targets.

'End of Pipe' Measures

The most widely used strategy to control emissions from motor vehicles consists of so-called 'end of pipe' measures. These include catalytic converters, cleaner, better maintained engines, and the alteration of oil-refining methods in order to produce cleaner petroleum products for use in motor vehicles.

Controls on motor car emissions have been gradually strengthened since the 1960s. Car engines have been redesigned so that they burn fuel more completely. In the US catalytic converters have been mandatory for cars since 1983, a rule that has applied to the EU only since the beginning of 1993.

Three-way catalytic converters, which when fitted during construction add relatively little to the selling price, convert NOx, VOCs and carbon monoxide to more inert substances. Because diesel engines burn fuel in an air-rich environment it has, in the past, been difficult to fit converters to them that will take the oxygen away from the NOx to produce inert nitrogen. Thus diesel-powered vehicles tend to produce more NOx than gasoline cars fitted with catalytic converters. In the future this problem is likely to be overcome as 'clean diesel engine' technology improves. Diesels also produce more particulates than cars powered by petrol, although the carbon monoxide, hydrocarbon and carbon dioxide emissions are likely to be lower than from petrol-driven engines. Heavy goods vehicles are, these days, mostly powered by diesel engines.

Catalytic converters reduce emissions by around 90 per cent, but only after the exhaust pipes have warmed up. It takes a couple of minutes for this to happen. According to a French survey, 30 per cent of car trips in cities are shorter than one kilometre.[2] Thus the so-called 'cold start problem' means that the reductions in emissions produced by catalytic converters are not quite as dramatic as the claims in the advertisements suggest. Nevertheless, because of them new European cars will produce much lower levels of NOx, VOCs and CO than old types of cars. The 'cold start' period is gradually being reduced.

In the US so-called reformulated petrol, which contains reduced amounts of VOCs, has, according to the 1990 Clean Air Act, to be sold in the nine smoggiest areas of the US. Cleaner, lower-sulphur diesel fuel is coming on the market.

In the 1980s environmentalists waged a campaign to remove lead from petrol. By the end of 1993 more than half of petrol sales were unleaded, even in the UK, and sale of leaded petrol is banned in the US after 1995.

Table 8.4 shows the proposals for upgrading emission standards for cars in the European Union. These will be broadly similar to US standards. It has been estimated that the new European standards will reduce 1990 levels of carbon monoxide, hydrocarbon and nitrogen oxide emissions by, respectively, 43 per cent, 60 per cent and 64 per cent by the year 2003.[3] It should be stressed that emission standards need to be continually updated to keep pace with traffic growth. Otherwise the emissions will begin once again to rise after 2003, barring radical improvements in vehicle fuel efficiency.

In the US reductions in some emissions will continue, although they will not be as dramatic as in Europe because most US cars are already fitted with catalytic converters. In some US states, including California, car manufacturers will have to sell an increasing proportion of so-called low emission and zero emission cars as time goes on. Northeastern states such as Massachusetts and New York have adopted programmes similar to California's and others are set to follow. California's 'low emission' limits are broadly similar to European targets for the year 2000 (see Table 8.4).

Table 8.4 Proposals for Upgrading 1993 EU Vehicle Emission
Standards

Standards taking effect from beginning of 1993			
		gm/km	
	CO	HC and NOx	Part.
All cars	3.16	1.13	0.18

Proposals to take effect at beginning of 1997			
		gm/km	
	CO	HC and NOx	Part.
Petrol	2.2	0.5	–
Indirect injection diesel	1.0	0.7	0.08
Direct injection diesel	1.0	0.9	0.10

Proposals to take effect from 1999			
		gm/km	
	CO	HC and NOx	Part.
Petrol	1.5	0.2	–
Diesel	0.5	0.2–0.5	0.04

Source: Warren Springs Laboratory, Stevenage

However, the shortcomings of reliance on tailpipe emissions controls are clear.

Despite some types of improvement, photochemical smog is still a problem in places like Los Angeles and Athens where geography tends to concentrate the pollutants. Ozone limits are exceeded in Los Angeles on around 150 days a year and on almost all days a year in Mexico City. Like all cities in developing countries, Mexico City still has far more lax environmental regulations than Los Angeles. The problem of NOx emissions is only being constrained in those parts of the industrialised world where regulations are being upgraded. Moreover, current regulations do not address the problems of PM10 emissions which come from all petroleum-driven engines, diesels being the worst offenders.

There seems to be an inexorable rise in transport-related carbon dioxide emissions. Transport tends to be the fastest rising contributor to carbon dioxide emissions of all the various sectors of most developed and of many developing economies.

In addition to this, most countries are increasing their dependence on foreign oil supplies in order to provide the increasing amounts of fuel needed to drive motor vehicles. This is less of a problem for oil-importing countries when oil prices are low but they cannot be relied upon to stay low for ever.

Clearly, if the pollution reduction targets mentioned earlier are to be achieved we are going to need more than 'end of pipe' measures, crucial though they may be.

Energy-efficient Vehicles

Much attention was focused by the US government in September 1993 on the possibility of producing super-lightweight vehicles that would need far less fuel to power them than current commercial designs. It was announced that federal funding would be going to US motor vehicle manufacturers to research fuel-efficient vehicles, as well as to research into fuel cells and more efficient gasoline engines.

Despite the considerable improvements in the fuel efficiency of US cars made since the mid-1970s, they are still more inefficient than the average European car, as can be seen in Figure 8.3.

Super-lightweight motor vehicles already exist in the form of motor racing cars. Motor racing cars are made of carbon fibre. This is very tough but many times lighter than the steel used to make conventional motor vehicles. Carbon fibre is also expensive to produce. However, in a much quoted paper produced in the summer of 1993, Amory Lovins claimed that in mass production such cars may reduce rather than increase production costs.[4] Lovins said that super-lightweight cars, when combined with other refinements, could, in the relatively near term, improve the fuel efficiency of the average US or UK motor vehicle fivefold and tenfold in the longer term.

Lightweight cars, produced from relatively short run 'epoxy dies', would be much more marketable as the number of model variations could be enlarged to suit personal taste more precisely. The car would consist of much fewer parts than the steel cars, thus cutting costs. Lighter materials are already being used to build hulls for boats.

By contrast steel cars need very long production runs and considerable investment is required in each model run. The technology is inflexible.

Two other factors identified by Lovins as ones that would improve fuel efficiency are reducing aerodynamic drag through better design and reducing rolling resistance through better tyres.

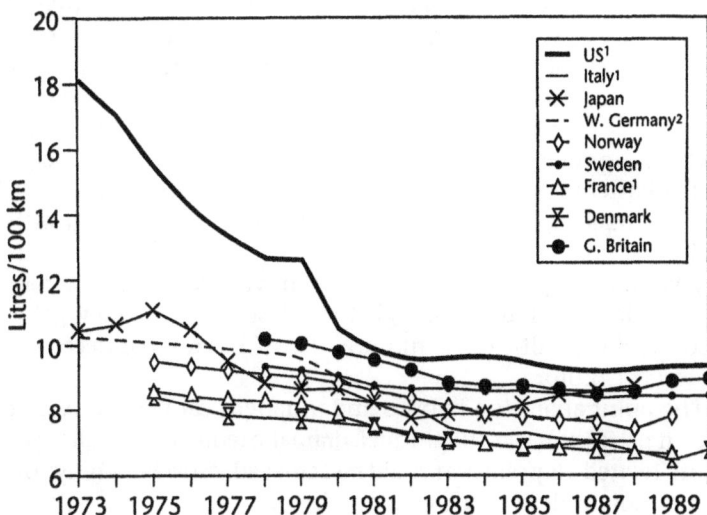

Figure 8.3 New Car Fuel Efficiency
(Sales-weighted Test Values)

1. Includes diesel; US includes light trucks
2. Excludes diesels

Source: L. Schipper, R. Steiner, M.J. Figueroa and K. Dolan, *Fuel Prices,
Automobile Fuel Economy and Fuel Use for Land Travel* (Berkeley, Ca:
Lawrence Berkeley Laboratory, 1993)

Both these techniques can be incorporated into designs of heavier
cars as well.

General Motors' 'Impact' prototype electric battery model is very
heavy because of the batteries, but its sleek design reduces
aerodynamic drag considerably. Pirella Armstrong, a tyre company
in California, is developing a tyre with 30 per cent lower rolling
resistance and 15 per cent less weight than standard tyres.

So-called hybrid electric propulsion systems may also help to
improve motor vehicle fuel efficiency. These have electric motors
but run on conventional fuels. Thus the benefits of running on
a cleaner, more efficient electric motor in urban areas and on an
internal combustion engine (ICE) during inter-urban trips (where
it performs best) can be realised without the need for large
amounts of heavy batteries.

Hybrid electric cars, such as the VW Golf diesel-electric hybrid model, use a small internal combustion engine and a small battery both of which are connected, in the case of a series hybrid, to an electric motor. The 'buffer' battery is recharged when the ICE is running. Regenerative braking wherein the brakes reclaim the energy otherwise lost during braking provide energy to the battery for storage. Switched reluctance motors are likely to bring major gains in electrical efficiency when combined with series hybrids.

There are also parallel hybrids where the vehicle can switch from ICE to electric mode, although these lose the efficiency gains associated with all-electric running. The ICE is not connected to an electric motor.

The point about the hybrids and ultralight car designs is that even if petroleum was still the fuel, dramatic reductions in all types of car-related air pollution could be achieved, especially if hybrid and ultralight designs are combined.

However long it takes before the 'super-lightweight' and hybrid cars are available on the mass market, we should not lose sight of the fact that there are already many designs of low-mileage cars available today. For example, the lightweight diesel-engined Citroën AX prototype has clocked up an average fuel efficiency of 5 litres per 100 km, only half as much as is used by the average new US car. Even greater improvements are possible with current technology.[5]

Regulations, such as the US Corporate Average Fuel Economy System (CAFE) can be used to improve motor vehicle fuel efficiency. The CAFE scheme was begun in the 1970s after the first oil crisis. It mandates manufacturers to ensure that their fleets conform to a certain average fuel economy, although its fuel economy regulations have not been significantly improved since 1986. The manufacturers do not like the system since they have to sell less popular brands and the decline in oil prices lessened pressure on them.

There has been controversy over whether high oil prices in the 1973–85 period would have improved fuel efficiency without CAFE standards. However, according to Lee Schipper et al., the fuel efficiency standards improved fuel efficiency much more than would have been the case without them. In fact, by the end of the 1980s fuel prices were no higher in the 1970s, and, taking the effect of improving US motor car fuel efficiencies into account, the cost of fuel per mile travelled was actually much lower in 1990 than it was in 1970.[6]

Some critics of regulatory attempts to improve motor vehicle fuel efficiency claim that motorists use the cost-cutting advantages of fuel-efficient vehicles simply to buy more fuel with the money they have saved. Some fuel savings are cancelled out by increased mileage, but research into motorists' behaviour indicates that in Europe about 70 per cent and in the US about 85 per cent of fuel efficiency gains are turned into real fuel savings.[7]

Taxes

Higher fuel taxes can also encourage fuel efficiency. Denmark and Italy (both heavily dependent on imported oil) increased their fuel taxes during the 1980s and fuel efficiency improved. In Italy small cars are further encouraged by a 38 per cent purchase tax (VAT) on large cars compared with only 19 per cent on smaller models. Average motor vehicle fuel efficiency is more than a third better than in the US. In the US gasoline taxes are only around a quarter the size of average motor vehicle fuel taxes in Western Europe. Fuel taxes make up 70 per cent of the cost of Western European petrol.

Although fuel tax increases are a much discussed (and often unpopular) option, less attention has been given to other ways in which fuel-efficient motor vehicles can be promoted, such as giving tax advantages to smaller cars (as in Italy) and in giving incentives for cars that are more fuel-efficient, whatever their engine capacity.

A 'purchase tax' or 'fee' could be levied on relatively energy-inefficient cars which can finance 'rebates' for more fuel-efficient vehicles. This would have the advantage of not penalising existing car drivers and giving everyone a choice. This is called a 'feebate' system. Although legislative proposals for a 'feebate' system have been put forward in California and Canada, the system is yet to be put into practice. It would represent a very low cost means of reducing emissions since greater fuel efficiency would mean lower fuel bills for the consumer.

Alternative Fuels

In recent years there have been many propositions for low-emission motor cars. These have included electric-battery cars, and cars run by natural gas, methanol, ethanol, and even hydrogen. Only electric and hydrogen-powered cars cut local pollution to almost zero. Electric cars may be 'zero emission' locally, but emissions will still be produced by fossil fuel power stations that produce the electricity in the first place. Similarly, emissions will be produced by fossil fuel energy used in the production of hydrogen fuel.

Ethanol produced from agricultural surpluses is providing a small contribution to fuel use but cannot provide a large amount without significant changes in land use and major subsidies. Methanol can be produced most cheaply from natural gas and it is being marketed with a gasoline blend in California. The advantage of methanol and ethanol over gasoline is declining as gasoline engines become cleaner, and Californians have lost a lot of their initial enthusiasm for these fuels. The idea of producing ethanol and methanol from biofuels such as wood is still popular among some environmentalists, partly because low carbon dioxide emissions would result.

Natural gas itself is also cleaner than gasoline. Natural Gas Vehicles (NGVs) run on compressed natural gas (CNG). At current prices they are a cheaper option than diesel or gasoline and further incentives through alterations in fuel taxes could be given. The gas industry points to expansion in demand for CNG as an alternative to diesel in several countries ranging from the US, The Netherlands and Australia to several states in Latin America.

NGVs are heavier than conventional vehicles because of the weight of the gas compression equipment. However, this penalty is not as severe in the case of battery-electric cars. Consequently, trucks can absorb the extra weight of CNG equipment with relative ease. Natural-gas-powered engines are thus a natural alternative to diesel engines which dominate the large vehicle market.

Compared to diesel-powered vehicles, NGVs produce much less of all the main emissions that cause local pollution, and they produce insignificant quantities of particulates, an increasing concern with diesels. In the long term, technical improvements to diesel engines are likely to narrow the gap between diesels and NGVs, but for the next few years NGV-powered trucks may be a useful stopgap measure for combating local pollution, although not global warming. Comparisons between the carbon dioxide produced by use of CNG and other fuels are shown in Figure 8.4.

Hydrogen cars are an interesting possibility, but there are still several problems to be overcome before they can be a mainstream competitor with other technologies. Hydrogen cars are quite heavy and there are no hydrogen pipelines to deliver the fuel. In the near term at least, hydrogen would still have to be produced from natural gas, making it expensive and an inferior economic proposition to using natural gas itself.

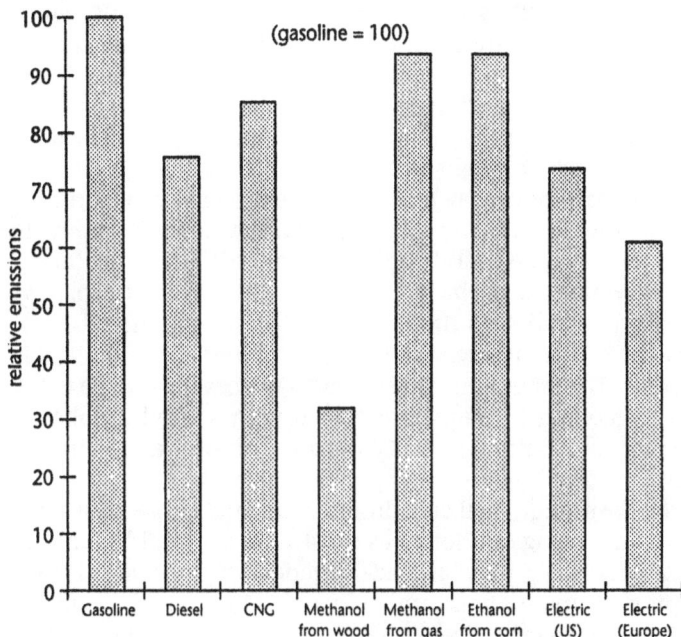

Figure 8.4 Proportionate Greenhouse Gas Emissions from
Different Fuels in Urban Driving[1]

1. These figures include the effects of car manufacture and fuel transportation.
Car manufacture accounts for about 10 per cent of the emissions using gasoline
as a fuel. Please note that the electric cars may only appear to have a relative
advantage because they have been built with lightweight materials and an
aerodynamic design, features which can just as easily be incorporated into
conventionally fuelled cars. It should be also noted that ethanol derived from
sugar cane grown in tropical zones will be associated with much lower
greenhouse emissions than are associated with ethanol grown from corn.

Source: adapted from figures given in *Cars and Climate Change*
(Paris: IEA/OECD, 1993), pp. 95–7

Hydrogen used as a motor vehicle fuel would produce little local
pollution, but, as is the case with electric cars, the amount of carbon
dioxide produced would depend on how the fuel was manufactured. Production from non-fossil sources would mean very low
carbon dioxide emissions, but hydrogen from non-fossil sources
is likely, in the short term, to be expensive.

Hydrogen use has also been associated with fuel cells, which promise efficient conversion of fuel into power. There are some prototype fuel cells in heavy vehicles marketed by Ballard. These systems use compressed hydrogen. Examples operate in buses in Vancouver. Such demonstration programmes could lead to innovative breakthroughs in the future.

Electric-battery cars at least have the advantage of a readily available fuel supply. The trouble is that they are very heavy. In the UK they are called milk floats. The weight of the batteries cannot be easily reduced except by replacing lead acid with more expensive battery types, and even then with only partial success.

Lead acid batteries need to be replaced every 25,000 miles at great cost. The maximum range in even the best prototype is 160 miles. Refuelling is long-winded even with a good number of refuelling points, although fuel costs using off-peak electricity will be low.

Sodium-sulphur, nickel-cadmium and nickel-metal-hydride batteries are being developed as substitutes for traditional lead-acid designs. Nickel-metal-hydride batteries, which are already used in laptop computers and cameras, would eliminate the need for replacements in cars. They improve the range to 250 miles and would allow much more rapid recharging.

The fundamental difficulty with any attempt by alternative fuels to compete with the petroleum-driven internal combustion engine is that oil is a very concentrated fuel, making it light to carry. Other fuels are usually less convenient to handle. Nevertheless, there is a fair head of steam behind efforts to improve prospects for alternatively fuelled motor vehicles. The California Air Resources Board has launched an accelerated programme of tightening emission controls with the added twist that they expect significant numbers of zero-emission cars. Car manufacturers are being mandated to ensure that 2 per cent of cars sold in California in 1998 are zero emission. This proportion rises to 10 per cent in 2003.

In effect the Californian quotas for zero-emission vehicles will be reached only by the electric-battery cars being cross-subsidised by people paying over the odds for conventional cars. Natural-gas-powered heavy vehicles may be the biggest beneficiary of any move towards alternative fuels on account of the cheap costs of natural gas. However, this may be only a quick fix while natural gas prices stay competitive.

It may be a while before solar photovoltaic cells (see Chapter 12) come down in price sufficiently to make solar motoring economic. When this happens it will occur in the context of very much lighter vehicles.

It can be seen from this analysis that in the medium term the technical changes to cars that will produce the biggest reductions in pollution are concerned with making them cleaner and more fuel-efficient, rather than running them on alternative fuels. That having been said, there is both a commercial and an environmental market in the near term for NGVs in the heavy vehicle and fleet vehicle sectors and also, in some limited areas, for electric-battery vehicles.

NGVs and electric-battery cars are often rejected because they make little impact on the global warming problem, but they can contribute towards the reduction of local, concentrated pollution. Asthma sufferers may be thankful for such benefits.

Reducing Traffic

Curbing the amount of traffic in general is the simplest way of curbing pollution from motor vehicles, at least in theory.

Traffic can be curbed by making travel more expensive. This can be done by putting up fuel taxes or by so-called road pricing. Road pricing schemes charge motorists for using the road. Schemes are in operation in Singapore and Oslo. Such schemes may reduce congestion, but of course they disadvantage people who are short of money. Improving public transport can partly offset such disadvantages in the cities, although not so much in rural areas where regular bus services are more difficult to organise.

A very cheap option is to ensure that towns and cities are planned so as to reduce the need to travel so much. The gains of such planning techniques are potentially bigger in the relatively dispersed urban areas in countries like the US and Australia rather than the more concentrated 'old' European cities. Nevertheless, such techniques may be important in stopping average car journey lengths becoming longer in European urban areas. The very densely populated country of The Netherlands is now putting great emphasis on 'environment friendly' urban planning policies, and considerable gains are expected. The developing countries could gain the most from such strategies since their cities are in the process of development.

Incentives can be given for inner-city developments, and planning permission can be withheld from out-of-town retailing and wholesale developments. Planners can ensure that 'high' rather than 'low' density housing is permitted, and that such development is sited near public transport routes. As with road pricing, this sort of strategy fits in with the idea of improving alternative transport modes. These include walking, cycling, buses and trains.

Changing Modes of Transport

Historically, urban systems have been planned by the road system rather than the other way around. When motorways, freeways or autobahns are built their exits serve as focal points for developments eager to secure communications advantages. Instead, cities could be planned around public transport networks.

Urban areas can be made much more friendly to walkers and cyclists with more pedestrianised areas and systematic cycleway systems. Streets can be made narrower and parking can be heavily restricted in town centres, making the environment less friendly towards cars. Parallel to this, more efficient bus and rail services can be delivered.

Increasing emphasis on alternative modes of transport can have significant gains in energy efficiency, as can be seen from Table 8.5. The energy efficiencies of trains, buses and coaches could, as in the case of cars, be improved if lighter materials are used to make the vehicles.

Table 8.5 Energy Requirements of Transport Modes

		Assumed load	Energy used (MJ) per passenger mile
Large car:	urban (20 mpg)	1.5	6.8
	highway (35 mpg)	1.5	3.9
Small car:	urban (35 mpg)	1.5	3.8
	highway (50 mpg)	1.5	2.7
Rail	commuter	65%	0.9
	intercity	50%	2.1
Bus		50%	1.0
Coach		65%	0.8
Motorcycle		1	4.3
Moped		1	1.4
Bicycle[1]		1	0.1
Walk		1	0.4

1. Although it is not covered here, an interesting variant on the bicycle is the 'human powered vehicle' which, since it is self-contained, has aerodynamic advantages over the traditional bicycle. It can go much faster and is even more fuel-efficient

Source: Various, reprinted from P. Hughes and S. Potter, *Routes to Stable Prosperity*, EERU Paper 61 (Milton Keynes: Open University, 1989)

Ridesharing

Besides shifting people on to different, more efficient modes of transport, the efficiency of car use could be improved by increasing the number of passengers per car. This is called 'ridesharing'.

The California Air Resources Board is promoting ridesharing schemes. Companies are encouraged to increase the number of employees travelling in vehicles with more than one person and car-pooling schemes where roads or lanes are designated for ridesharers only are being implemented.

Ridesharing schemes have been implemented in many places, including Japan and Italy. It has often occasioned jokes being made about the use of professional or even plastic inflatable passengers to beat the ridesharing rules. A more fundamental criticism is that highway authorities are using ridesharing to justify road widening and then simply opening up ridesharing lanes to all traffic. Such tricks have rather discredited this policy.

The Future

Transport problems, which include local air pollution, congestion, oil dependence, carbon dioxide emissions and the environmental damage caused by building roads are complex. They are not reducible to one or even two solutions. There are two extremes in the debate about solutions. One says that cleaner cars are the solution, while the other says that most motor vehicles would be unnecessary in a truly green society.

On the one hand there is a lot that can be done to plan towns and cities better and to improve alternative transport modes to lessen greatly the necessity for and attractiveness of motor cars. On the other hand, the flexibility and freedom that motor cars give to the consumer is too seductive for us to downgrade the great improvements that can be made in cleaning up motor cars and making them much more energy-efficient.

Thus, many analysts stress that a 'systems approach' is needed and a range of solutions ought to be implemented. A range of policy instruments is needed. If they are all implemented with vigour then a truly balanced transport strategy which heralds a future of much more energy-efficient, much cleaner motor vehicles, better planned towns and cities and efficient alternative transport modes can produce solutions that are sustainable in all environmental, economic and cultural senses. These policies can combine to produce radical absolute reductions in emissions of all sorts in the context of reduced rather than increased transport costs. The issue is whether environmentalists can mobilise behind such policies and ensure that they are implemented.

As in the case of other energy efficiency policies, such strategies may fit in with information-technology-induced changes such as the increasing practice of 'telecommuting' down modem lines which reduces the need to travel to and from work.

9

Cleaner Coal?

Coal is by far the most plentiful fossil fuel. However, where natural gas supplies are cheaply available, coal has been or is being pushed out of its once traditional domestic and industrial markets. It is also facing competition in the electricity sector from the advance of CCGTs and natural gas cogeneration.

This competition has engendered a great deal of industrial and political stress in the UK, Germany and the US as miners fight for their jobs. However, where natural gas supplies are readily available the economic and environmental advantages of natural-gas-burning technology are likely to triumph in the medium and long term.[1]

Despite these developments, coal is a long way from being written off as a major fuel. Indeed, environmentalists are concerned about the increasing size of the Chinese coal burn and the impact this will have on the level of carbon dioxide emissions. The Japanese are worrying about the acid emissions from Chinese coal-fired power stations. They may pay the Chinese to use cleaner coal-burning technology.

China, which is responsible for a quarter of all the world's coal production, regards coal as its prime energy resource, although increasing discoveries of Chinese natural gas reserves may mean that natural gas will supply much of the future growth in Chinese energy consumption. Energy- (and gas-) scarce Japan is greatly expanding its coal-fired power station capacity and so are several other countries.

Although it is likely that natural gas will overtake coal as the second biggest supplier of global energy needs, coal is likely to remain a major player for the foreseeable future. Electricity production is likely to be coal's last bastion.

It is a pretty large bastion. Coal produces 40 per cent of the world's electricity, the amount of electricity generated is increasing and its proportion of delivered energy is rising as the 'electricity intensive' service sector expands.

Given the mutual importance of coal and electricity I shall concentrate attention on technology that can clean coal in electricity production.

Coal is inherently a dirty fuel, although the level of dirtiness varies with the type of coal. Coal contains around 5–10 per cent ash (which gives rise to dust emissions), 1–3 per cent sulphur, 1–2 per cent nitrogen and various types of heavy metal. Anthracite coals have the highest energy content per tonne, and soft, brown coal, or lignite, the lowest. Bituminous coal, the most common, is in the middle.

Sulphur emissions can be cut by using low sulphur coal. This involves, as is the case with Britain, importing coal from places like Colombia, or in the case of the US using low sulphur deposits concentrated in Montana and Wyoming. Britain does have low sulphur coal seams, but most of the mines producing it have been closed down.

If burnt without any filtering devices, coal gives off large volumes of oxides of nitrogen and sulphur and particulates. Coalmining causes subsidence if done underground and disfigures the landscape if done through open cast or strip-mining. Coalmining has traditionally been extremely hazardous, although since the Second World War the number of deaths has been reduced in most OECD countries because of health and safety regulations. Health and safety is rather more lax in the case of big coal exporters such as Colombia and South Africa where wages and working conditions are generally low.

In most OECD countries the regulations governing operation of coal-fired power stations have been greatly tightened since the 1960s. Besides the requirements for control of NOx, SO_2, and dust, taller chimneys and cooling towers have been needed, the former to disperse pollutants better and the latter to ensure that fish will not be killed by discharges of warm water into rivers. Consequently the capital costs of coal-fired stations have gone up. According to one estimate, the cost of various sorts of additional equipment pushed up the capital costs of new plant by threefold in the period 1969 to 1981.[2]

However, in countries like China where low cost is important and environmental criteria are of relatively little importance, coal-fired power stations are built without such equipment.

Conventional coal-fired power stations burn pulverised coal. Dust particles can be trapped by electrostatic precipitators, although control of the larger particles that make visible smoke is rather more complete than control of the finer but still potentially harmful particulates.

NOx emissions can be cut by around 40 per cent by adding so-called low NOx burners. These consist of equipment which adds ammonia or other compounds into the boiler furnace which turns the NOx into water, nitrogen and carbon dioxide. Selective catalytic reduction (SCR) techniques can also be employed. These will reduce the NOx emissions still further, perhaps to 90 per cent, but SCR techniques are rather more expensive.

Flue gas desulphurising (FGD) equipment can be added to remove SO_2 emissions, generally with around 90 per cent effectiveness. FGD is most commonly achieved by using limestone to react with the sulphur to form gypsum. There is a complication in that very large quantities of gypsum are produced. Such quantities can be too large for building industries to absorb so that disposal sites have to be found. Large amounts of limestone have to be mined. This causes disturbance to the countryside.

There is an alternative FGD method using caustic soda as well as limestone, and this process uses limestone less voraciously. However, it tends to be more expensive and leaves sulphuric acid, liquid SO_2 and sulphur as by-products which themselves require disposal.

Since the 1970s, several designs that produce inherently lower emissions have evolved. They come under the general heading of 'advanced' coal-fired power stations.

The circulating fluidised bed combustor (CFBC) is like conventional designs except that the coal is burned in a bed of sand and ash which is constantly blown by a fast circulating stream of air. Limestone can be added to the bed, removing the need for FGD, and since the system runs at a relatively low temperature (about 500 degrees C), NOx emissions are relatively low.

However, so-called pressurised fluidised bed combustion (PFBC) seems to be able to achieve higher efficiencies. Coal is burned at high pressures and the hot coal combustion gases drive a turbine. The waste heat is then used to raise steam to drive a further turbine. It is a combined cycle operation achieving efficiencies above 40 per cent as opposed to around 38 per cent for a conventional plant (fitted with FGD) or a CFBC plant. Like the CFBC design, PFBC plant captures sulphur in the bed and the NOx emissions are low.

Integrated gasification combined cycle (IGCC) plant involves the total 'gasification' of the coal using oxygen which converts the coal into carbon monoxide and hydrogen. This is then driven through a combined cycle operation. IGCC plant can achieve efficiencies of 43–45 per cent and are especially effective at reducing SO_2 and NOx emissions; SO_2 emissions are reduced by around 95 per cent and NOx emissions by around 70 per cent compared to traditional coal-fired power stations. On the other hand, the Lurgi gasification system used for the IGCC design produces waste and toxic tars. These can be used as roadbuilding material.

The IGCC concept is widely tipped to be the leading clean coal technology of the future, although the majority of sets in operation in 1993 used high-sulphur oil refinery residues as the fuel. Orimulsion, a bitumen-based fuel from Venezuela, is also being used to fuel IGCC plant.

Other 'advanced coal' designs include the British Topping cycle which involves partial gasification and combustion of char in the steam cycle. The Germans are also developing advanced coal-fired power stations.

Traditional coal-fired power stations can be linked to cogeneration systems but they are relatively inefficient users of energy compared with gas-fired units and do not save much carbon dioxide emissions. Cogeneration systems based on advanced, gasification-based designs do rather better.[3] Nevertheless, the capital and operating costs will be much higher than for natural-gas-fired units, particularly so at smaller sizes, and they will not cut carbon dioxide emissions nearly as much as an efficient gas cogeneration system.

There are now a number of examples of CFBC, PFBC and IGCC plant operating. It is hoped that their quite high capital costs (around £1,000/kW) will be reduced with wider adoption.

Conventional power stations fitted with bolt-on FGD and low NOx burners may cost around £750 to £800 per kW, and thus have so far tended to be used rather than advanced coal designs. Further tightening of emission standards may change this. Electricity from advanced coal designs is unlikely to cost less than around 3.5 p/kWh given coal prices of £1.30/GJ, a 10 per cent discount rate and a 15-year contract.

As concern over the impact of carbon dioxide emissions has grown over recent years, so research studies have been conducted into ways of scrubbing carbon dioxide from coal and other fossil-fuel-fired plant and disposing of the gas in places from which it will not return to the atmosphere.

10

Carbon Sequestration

An obvious strategy to control carbon dioxide emissions from fossil fuel burning is to collect and trap the gas so that it does not increase carbon dioxide levels in the atmosphere. Called 'carbon sequestration', it is a strategy that has attracted an increasing amount of attention over the last few years. The most natural way of achieving it is to cultivate trees and other plants to absorb carbon dioxide.

Planting Trees

One of the causes of rising levels of atmospheric carbon dioxide is the destruction, by burning, of trees. The effect of burning trees can be neutral if forests are replanted. However, for the most part trees are being destroyed without being replaced. Tree-burning is said to account for very roughly 10 per cent of new additions to global warming.

Many are suggesting that Western power companies could subsidise the planting of trees in order to mop up carbon dioxide emitted by power stations. US and Dutch power utilities have paid for tree-planting schemes in places like Guatemala, Malaysia and Czechoslovakia.[1] The US government is considering the idea of offsetting US emissions by subsidising tree-planting schemes in other countries.

However, many ecologists criticise such 'offset' schemes. One limitation is that after a few decades (depending on the method used) a given forest will stop absorbing significant quantities of carbon dioxide. This means that the land area that has to be afforested needs to be continually expanded if a given rate of absorption of carbon dioxide emissions is to be achieved on an indefinite basis. For example, the UK's current tree-planting programme (30,000 to 40,000 hectares a year) is likely to absorb 1 to 2 per cent of fossil-fuel-related carbon dioxide emissions. If the UK tried to absorb just 10 per cent of its current production of carbon dioxide then the whole of the UK would be covered with trees in less than a century![2]

Another problem is that if trees are planted in developing countries, the effect of such plantations could simply be to make peasants chop down trees elsewhere to plant their crops. The general message from environmental groups is that Western states ought to be funding tree-planting or tree-preservation schemes for their own worth, not as offsets to allow them to continue their own carbon dioxide emitting profligacy.

Tree-planting schemes are important for a number of ecological reasons but it is very unlikely that they can offset more than a small portion of carbon dioxide emissions. A variant on the tree-planting scheme is to grow biomass for energy purposes. Many regard this as a more efficient method of carbon sequestration (see Chapter 12).

Decarbonisation

Another method of carbon sequestration has been called 'decarbonisation'. This involves trapping carbon dioxide emissions from power stations and disposing of them in a place from which the gas will not seep back into the atmosphere, or at least not for a very long time. This has often been regarded as an oddball proposition. However, research sponsored by the International Energy Agency (IEA) has made the topic more respectable.

Some carbon dioxide is already recovered from fossil fuel power stations and pumped into the ground into partly empty oil wells in order to effect so-called enhanced oil recovery (EOR) in the US. Natural gas is sometimes pumped back into exhausted fields to provide stores to satisfy peak demand. In principle, therefore, there seems no reason why carbon dioxide could not be removed from power stations and pumped down disused oil wells or gas fields where the natural geological traps should ensure that it stays put.

There are various methods by which carbon dioxide could be removed, the current method being to use amine solutions to dissolve it. This involves a 30 per cent loss in a power station's efficiency. However, work by British Coal suggests that if an IGCC coal plant is used in conjunction with a membrane separation system, the efficiency losses could be reduced to about 17 per cent.

Provided the oil and gas wells (where the carbon dioxide would be placed) are within 100–200 km from the power station, the cost of the pipelines needed to take the carbon dioxide to its disposal wells might not be too prohibitive. Altogether the electricity generation cost could be around 5.5 p/kWh at a 10 per cent discount rate.[3] However, many of the potential disposal sites would be rather farther away, thus increasing costs further.

The Japanese have shown interest in disposing of carbon dioxide on the ocean floor. There are deep ocean trenches close to Japan. In fact, the technology that could pump carbon dioxide down to the ocean floor (where it would stay as a liquid at the high pressures prevalent at such depths) has not yet been developed, although solid blocks of carbon dioxide could be tipped over deep trenches.

Near-term solutions could involve dissolving the carbon dioxide at lesser depths, the downside being that a large proportion of the carbon dioxide would re-enter the atmosphere in a few hundred years. However, this does not necessarily rule out the option since this would allow time for alternatives to fossil fuels to be developed, thus limiting climatic impact. In many cases the length of pipelines needed to take the carbon dioxide to oceans would make the option very expensive.

Other possibilities include disposing of the carbon dioxide in deep underground aquifers. This has been researched and discussed in The Netherlands and Denmark.

There are considerable environmental controversies about ocean disposal and disposal in aquifers. Many ecologists would complain of the possible damage to sea-life. In Denmark many have expressed fears that carbon dioxide injections into aquifers could lead to dangerous gas escapes (the gas will suffocate life in large quantities), groundwater pollution, ground instability and other problems.

Going further up the Richter scale of zany ideas, researchers at the University of Stuttgart have suggested storing the carbon dioxide 'as giant insulated blocks of dry ice'[4] on the land surface. The 400-metre-wide blocks of dry ice would take thousands of years to evaporate completely. (I suspect there might be one or two planning objections to this sort of scheme.)

The potentials for carbon dioxide storage are shown in Table 10.1. Realistically it seems likely that if carbon dioxide was to be scrubbed from power stations, disposal in disused oil and gas wells would be the preferred option. Even though this is a much more limited option than ocean disposal, it would still offer many countries the option of disposing of large proportions of their carbon dioxide production from power stations for a few decades.

Table 10. 1 Potential Carbon Dioxide Storage Capacity (GT of CO_2)

Ocean disposal	70 million
Terrestrial	high
Aquifers	320
Exhausted gas wells	320
Exhausted oil wells	150
Enhanced oil recovery	15
Global forest management	180–370

Note: current global annual CO_2 production from fossil fuel sources is around 22 GT
Source: *ENDS Report 219*, April 1993

The question of whether this is ever likely to be taken up is a different matter. Society would have to be both very concerned about global warming and very short of other options before it would decide to accept increases in the generation cost of fossil fuel electricity of 40 to 50 per cent just for the sake of storing carbon dioxide. It seems highly unlikely that countries like China, which at the moment is not keen even to cut sulphur and dust pollution that clogs up its children's lungs, would go to such trouble.

11

Nuclear Power

The Promise of Nuclear Power

In many ways the promise of civil nuclear power has not changed since 1942 when Fermi demonstrated the first sustained nuclear reaction.

Its promise lies in its ability to substitute for fossil fuels thus avoiding the problems that go with them. These include dependence on imported fossil fuels, fear of physical depletion of fossil fuels, and their environmental hazards, especially air pollution.

Resources of uranium, which provides the fuel for conventional nuclear reactors, are not infinitely abundant. However, the nuclear strategy that evolved in the 1950s and 1960s envisaged that this problem could be largely overcome by recycling the uranium.

Naturally occurring uranium contains two isotopes, uranium 238 (U238) and uranium 235 (U235). These are the same element, but each has a slightly different number of neutrons. U235 makes up only 0.7 per cent of the total mass of naturally occurring uranium, the rest being U238.

However, the U235 is fissile, meaning that a process of radioactive decay occurs in the natural state. Individual atoms 'fission' or split up into smaller atoms, giving off heat and other forms of radiation including neutrons. Under the right conditions these nuclear particles can cause further fissions in other U235 atoms, so that a self-sustaining chain reaction can be established. U235 is the active ingredient in conventional nuclear reactors.

In many types of reactor, including the most common version, the pressurised water reactor (PWR), the proportion of U235 is enhanced through an enrichment process. As time goes on the proportion of U235 in the fuel rod declines and the amount of impurities increase. The fuel rod has to be replaced.

The spent fuel rods can then be reprocessed. U238 can be recovered, as can plutonium. These elements can be used in another type of reactor called fast breeder reactors. They are called 'fast' because the neutrons that cause the nuclear reactions are not slowed down by a 'moderating' substance as is the case with conventional reactors. The fast neutrons from decaying plutonium turn the U238 into plutonium. The reactor creates or 'breeds' its own fuel. Thus, the use of uranium resources could, in theory, be multiplied by up to a hundred times.

Then there is nuclear fusion. While existing nuclear reactors use fission -- the splitting of atoms of heavy elements to release energy -- nuclear fusion involves the fusion of light atoms, in particular hydrogen, to release energy. Given the abundance of the materials needed for nuclear fusion, the potential of fusion power is extremely large.

Such is the promise of nuclear power. What, to date, has been the reality?

The Progress of Nuclear Power

The world nuclear power programme was launched by Eisenhower's 'Atoms for Peace' campaign in 1953. In the mid-1950s the first (small) quantities of energy from nuclear reactors was fed into electricity networks. International organisations were set up to promote nuclear power. 'Euratom' was set up under the Treaty of Rome in 1957 as a key part of the newly formed European Community.

The depletion of fossil fuel reserves, fear of increases in fossil fuel prices and concerns about the security of Middle Eastern oil, and the demonstrably enormous power of the atom meant that nuclear power was thought by most people to be synonymous with 'progress'.

Nuclear power began to take off in the US in the second half of the 1960s. This followed the famous 'Oyster Creek' contract, awarded in 1964 and proclaimed to be the first in a run of orders for power stations that produced cheaper electricity than coal-fired power stations. By 1973 nearly 200 nuclear power stations had been ordered in the US, several sets were already in operation in the UK and big programmes were underway in Japan and France. Yet even before the oil crisis there were signs of trouble.

The capital costs of building the power stations were much higher than expected. The General Electric/Westinghouse consortium that had organised a series of 'turnkey' contracts for PWRs in the US reduced their involvement after severe financial losses. A lot of the problems were environmental in origin. People were demanding that safety measures be upgraded. In addition the complexities involved in the technology and the problems that went with scaling up from small to larger 600 MW and 1000 MW sets proved much more difficult to deal with than was first expected.

Then came the oil crisis. Far from helping nuclear power the high energy prices depressed the demand for electricity. Utilities slashed their power station building plans. The most expensive power projects, which usually turned out to be nuclear, were the first to receive the chop. Worldwide, around two-thirds of the nuclear power stations ordered before the oil crisis were later cancelled.[1] In the US no nuclear power station ordered after 1974 has ever been completed.

Some countries expanded their nuclear power programmes much more smoothly than in the US, and, initially at least, with lower costs. These countries included France, the Soviet Union, Japan and latterly South Korea.

The French nuclear programme is the 'jewel in the crown' of the nuclear industry. Although much smaller in reactor numbers than the US programme, French nuclear power was supplying 73 per cent of French electricity by the end of 1993. The French programme was started by De Gaulle in 1969 in order to reduce French dependence on imported oil. France has few of its own fossil fuel resources.

Other countries encountered a series of problems. In the UK the first series of reactors, the Magnox stations, had to be downrated because of safety problems. They were relatively expensive to operate. A new advanced gas-cooled reactor (AGR) series was planned, but by the early 1970s this had run into major problems.

A series of nuclear projects in developing countries, including Argentina, Brazil, the Philippines, Egypt and Iran, proved to be very expensive disasters.

Elsewhere, especially in the West, there have been political problems. In some countries (such as Sweden, Austria, Italy and Denmark) referendums have stopped nuclear power stations being built. In several countries, including Sweden and Germany, there are strong pressures (so far resisted) for currently operating nuclear power stations to be closed.

By 1993 there were about 400 nuclear power stations working worldwide (around 100 in the US) producing around 7 per cent of the world's primary energy supply and about 17 per cent of all electricity. However, expansion in the West has slowed almost to a halt. The 1979 Three Mile Island reactor meltdown and the 1986 Chernobyl disaster, along with intense controversy about nuclear waste made it seem that environmental and political factors had damaged nuclear construction programmes. Closer examination of the reasons for cancellations of orders reveals that commercial and economic factors were very often, if not usually, the direct cause.

The second stage of the nuclear strategy, fast breeder reactors, has run into operating problems. A few demonstration reactors have been built in France, the Soviet Union, Japan and the UK (the UK's fast reactor at Dounreay has been closed down), but only one new Russian plant seems likely to go ahead in the near future. With a glut in the supply of uranium (made ever more acute by the spare material from dismantled Russian nuclear warheads), there is no demand for the technology. Fast breeders have proved difficult to operate (they use highly reactive liquid sodium as a coolant), expensive and even more prone to safety fears than ordinary reactors since they could, in theoretical worst-case scenarios, produce an actual nuclear explosion.

The picture for the nuclear industry is not one of undispelled gloom, for although the nuclear power station building programme has all but stopped in the West and failed to make any progress in developing countries, there is a continued trickle of orders in former communist countries and Eastern states such as China, Japan and South Korea.

The factor that most of these states have in common is shortage of energy. The collapse of the Russian oil industry has meant that the country has very little to spare for domestic consumption; it needs the hard currency earned from oil exports. In December 1993 the Russian government announced a plan to build 30 reactors, although whether they will all be built remains to be seen. Countries like Korea and Japan have few indigenous energy resources. Even Ukraine, which suffered terrible contamination from the Chernobyl accident, faces an energy crisis of such extremes that it has been driven to reverse its decision to close the Chernobyl-type RBMK reactors. The Ukrainians have decided to complete construction of reactors which were 'frozen' in 1986.

Even in the West there appears to be a couple of rays of hope, paradoxically for ecological reasons. 'Can global warming save nuclear power?', is the question being asked by several analysts. In order to try to answer this question I shall analyse the main problems faced by the nuclear power industry.

Problems in the Nuclear Power Industry

Nuclear Safety

Since nuclear power relies on nuclear reactions which are difficult to control, there are bound to be fears about its safety. Reactors need to maintain just enough neutron activity to keep the reactor working but not so much that the nuclear reaction runs out of control. Conventional nuclear reactors (often misleadingly called thermal reactors) cannot produce nuclear explosions, but if runaway nuclear reactions occur, and if cooling systems fail, the resulting heat can melt down the fuel rods and trigger violent explosions. This is what threatened to happen at Three Mile Island and is what actually happened at Chernobyl.

There are wide variations in the estimates of deaths resulting from the Chernobyl accident, but the World Health Organisation estimated, after a survey of cases, that there could be 40,000 cases of child thyroid cancer afflicting people living within a 300 km radius of Chernobyl.[2]

It may rightly be said that other energy sources sometimes present serious safety problems. LNG (Liquified Natural Gas) facilities are a case in point. But then there is great controversy over LNG storage facilities, as can be seen in the problems developers have had in securing planning permission in Italy for such plant.

The demands for greater nuclear safety will not be forgotten. Since Chernobyl there has been talk of reducing the theoretical risk of major reactor accidents to one in 10 million per reactor year, a target which reactors in Western Europe do not meet.[3] There are proposals for 'inherently safe' reactors employing 'passive' safety features (that is, they will not blow up whatever foolish Chernobyl-type technicians try to do with them). However, they are developments of already existing designs.

Some challenge the notion of inherent safety.[4] I do not doubt that nuclear safety can be improved, but the question is, at what financial cost?

Nuclear Costs

As was explained in Chapter 5, the biggest elements in the costs of electricity from nuclear power stations are the capital costs, the costs of building them. If people demand increased safety then this has an upward pressure on capital costs. The trend has been for higher and higher safety standards. This upward pressure on costs is not good news for nuclear power because the costs of technological competitors such as gas turbines have been coming down.

Since nuclear power stations take from six to eight years to build, a lot of interest accrues on money spent during the construction period. At a 10 per cent discount rate this adds around 50 per cent to the capital costs.

Various studies[5] have shown that the costs of nuclear power have risen since the 1960s, and may be considerably higher than official estimates.

MacKerron records that in the US, after taking account of inflation, the cost of the later PWRs was 3.6 times the cost of the early models.[6] In France, capital costs have doubled since the earlier models. Cracks in reactor vessels during the 1980s have also added greatly to the cost of maintaining the power stations, making French nuclear power more expensive than its image suggests.[7]

The relative cheapness of the French programme compared with others stems partly from the economies of scale allowed by the massive nature of the French nuclear programme. This involved the wholesale conversion of the electricity generating system, the scrapping of existing power stations with relatively low avoided costs and a decision to ignore opportunities for DSM and cogeneration.

Even in South Korea, where the nuclear programme has often been acclaimed as what is possible in the developing world, the capital cost of building a nuclear power station has almost doubled in recent years.[8]

The UK programme has been spectacularly dogged by misfortune. The latest project, at Sizewell B, seems likely to produce power at something like 7–9 p/kWh.[9] The capital costs including interest charges are, at £4,000 per kW, around eight times the capital cost of gas-fired power plant. In 1994 Nuclear Electric claimed that a new proposal for a twin reactor at Sizewell could cost 3.7 p/kWh, but this estimate was based on the expectation that capital costs would be cut by nearly half compared with Sizewell B and that plant availability would be very high. Given these factors, it is difficult to take this costing seriously.

Estimates for the costs of decommissioning nuclear power stations are increasing. The cheapest short-term solution is to 'entomb' the nuclear facilities for at least several decades, and leave the task of dismantling the most seriously irradiated parts of the plant to future generations. This may not please people who live close to the radioactive hulks. On the other hand, even complete decommissioning still leaves the problem of what is to be done with the radioactive waste materials that are left over after the power plants have been dismantled.

Many believe that we should at least set aside sufficient funds now for future generations to complete the task of decommissioning. According to Krause et al.[10] the cost of decommissioning PWRs could, in their highest estimates, contribute up to around 20 per cent of the costs of supplying electricity from French PWRs.

In theory, decommissioning can be dealt with by a sinking fund which accumulates interest over time, although in practice some funds have either lost money or have not been established at all. According to a report in the *Wall Street Journal*[11] US utilities should have so far put aside $33 billion, but in fact have put aside only $4 billion.

Whatever decisions are made about decommissioning, there seems little chance that nuclear power will be competitive with power from fossil fuel plant. It should be emphasised that although decommissioning costs have, in recent years, been widely cited as the reason for the high cost of nuclear power, the reality is that electricity from new nuclear power stations is very expensive because of high construction costs and the length of time taken to build the power stations. Given the tightening of safety regulations and the consequent escalation of costs of building new nuclear plant it seems unlikely that electricity from new US nuclear power stations (and orders seemed remote in 1993) would cost less than around 9 c/kWh.

Some examples of costs of electricity from new PWRs are given in Table 11.1. They seem to cluster above the 5 p/kWh mark. In the French and German cases Krause et al. say that the official figures do not include various items like insurance costs and that they do not make adequate allowance for decommissioning costs or increases in maintenance costs resulting from premature ageing of the installation. The US, UK and Korean estimates in Table 11.1 make only very small allowances for decommissioning and no allowance for insurance.

A pattern has emerged over the past 40 years where utilities and government agencies project construction costs which invariably prove to be large underestimates (sometimes by factors of two or three times), capacity factors are optimistically high, and low discount rates are used. Often interest during construction is omitted entirely, and insurance costs are invariably borne by the state. Commercial energy projects could not be executed on this basis.

Table 11.1 Costs of Electricity from New PWRs[1]

	Germany	France	UK	Korea	US
Capacity factor	65%	65%	70%	80%	70%
Discount rate	5%	7.5%	10%	8%	7%
Contract length (yrs)	23	23	20	20	20
Cost (p/kWh)	5–6.7	3.2–5.2	7–9.0	5.0	6.0

1. Costs have been converted into 1993 prices except for Korea which is in 1991 prices

Sources: France and Germany: F. Krause et al., *The Cost of Nuclear Power in Western Europe*, Vol. 2, Part 3D of *Energy Policy in the Greenhouse* (El Cerrito, CA: International Project for Sustainable Energy Paths, 1994), taking median figure of their estimates as the upper figure and the official government figure as the lower estimate; UK: evidence (from MacKerron) accepted by UK Select Committee on Energy in, *The Costs of Nuclear Power* (London: HMSO, 1990); Korea: Chung-Taek Park, 'The Experience of Nuclear Power Development in the Republic of Korea', *Energy Policy*, August 1992; US: MacKerron 'Nuclear Costs, Why Do They Keep Rising?', *Energy Policy*, July 1992. US and Korean costs include typical assumptions about interest during construction and assume that operation and maintenance, fuel cycle and decommissioning costs total around 1.5 p/kWh

It is not surprising that there are no nuclear power stations being built wholly from private money. The lack of state sponsorship (other than responsibility for waste disposal) in the US killed off their nuclear expansion programme.

There seems little chance of new nuclear projects in the US. There are question marks about how long the older nuclear power stations can be kept going. An increasing number need expensive retrofits and it seems unlikely that many will reach the end of their 40-year licences, never mind continue afterwards, without federal assistance.[12]

In 1989, when the UK power generation sector was privatised, most of the UK's PWR programme was cancelled. National Power feared that the PWRs would sink their share flotation.

The nuclear power industry was kept in the public sector though a publicly owned organisation called Nuclear Electric which, supported by a subsidy from electricity consumers called the fossil fuel levy, continued to build another power station at

Sizewell B. Nuclear Electric want to build a further twin reactor at Sizewell C, but they will need a very large public subsidy to allow them to do this.

One good sign for nuclear power in recent years has been the steady improvement in performance of existing nuclear power stations. Capacity factors for PWRs increased from 57 per cent in 1980 to 73 per cent in the year ending June 1993.[13]

The avoidable costs of electricity from existing power stations (that is, the costs saved by not running them at all) are low in many cases, perhaps 1 to 1.5 p/kWh. This is because capital costs have already been committed. Of course as the reactors age they need repairs which lead to increasing costs. In the UK those Magnox reactors still in operation have running costs of well over 4 p/kWh,[14] but they are kept going by state subsidies. Anti-nuclear campaigners claim that these power stations emit much higher levels of radioactivity than would be permitted for new nuclear plant.

The biggest problems, in cost terms, for nuclear power come with new projects and projects coming to the end of their lives. New projects have very high capital (mostly construction) costs. As the number of 'retired' nuclear installations increases, so do the arguments about what to do with them and whether the costs of fully decommissioning them should simply be passed on to future generations.

All this is bad news. Even if capital cost increases can be stabilised and high capacity factors achieved, recent experience suggests that the cost of electricity from new nuclear power stations will often be a lot more than 5 p/kWh (1993 prices) in countries where there are high safety standards.

Nuclear Waste

Nuclear waste is created in two basic ways. It is created from the equipment used in nuclear installations, ranging from protective clothing to the plant (after it has stopped generating electricity).

Nuclear waste is also created by the spent nuclear fuel rods themselves. According to 'classical' nuclear theory, the fuel rods are 'reprocessed' (at places like Sellafield in Cumbria in the UK) to recover the unused U238 and the plutonium which can both be used to fuel fast breeder reactors. This leaves a large amount of nuclear waste consisting of the radioactive impurities contained within spent fuel rods. Something has to be done with this intermediate and high-level nuclear waste.

The issue of nuclear waste is seen as a technical matter by the nuclear industry, but their opponents have elevated it to the level of a moral principle. In 1973 E.F. Schumacher in his now classic *Small is Beautiful* text described the accumulation of unrecyclable toxic nuclear wastes as a transgression against life itself and a worse crime than any perpetrated by man.[15]

The nuclear waste problem has still not been resolved. No country in the West has managed to secure a permanent site for depositing high-level nuclear waste. The US government hopes that the Waste Isolation Pilot Plant (WIPP) in a desert area of New Mexico will be the world's first home for half a century's worth of accumulated US nuclear waste. However, legal wrangles continue. Proposals for waste disposal sites in all the major Western nuclear states have run into fierce local and regional opposition.

The problem is that the waste will remain radioactive for thousands of years. Although the nuclear industry has argued that deep waste repositories can be built strong enough to stop radioactive leakage, others doubt whether the site can be monitored for thousands of years (the half-life of plutonium is 24,000 years) and whether anybody can guarantee that geological conditions will remain the same over such long periods.

In some conditions nuclear waste can cause serious accidents, which is what happened at Chelyabinsk in Russia in 1957 in an incident that has still not been given a full official explanation. Nuclear waste contamination in the former Soviet Union has reached nightmarish proportions. A 40 hectare lake near Karachay, around 1,450 km east of Moscow, was used as a dump for nuclear waste (resulting from production of 'military' plutonium) for nearly 40 years. By 1990 it contained two and a half times as much radiation as was released by the Chernobyl accident.[16]

Radioactive contamination at the US's Hanford military weapons reservation near the Columbia River in northwestern US is so bad that the area has to be quarantined and nobody knows when the site can be reclaimed.

The contamination caused by reprocessing facilities such as those at Sellafield is the subject of great controversy. There is continuing debate over whether excess leukaemia cases around the reprocessing plant are caused by the nuclear facility.

There have been suggestions that the waste could be fired off into outer space, or into the sun. However, given the fallibility of rocket technology the problems with 'deep space' disposal may be just as severe as those which go with disposal underground. Disposal at sea has been banned under an international convention.

A further problem with nuclear waste is the plutonium that is separated from the rest of the waste. The fast breeder reactors that the plutonium was supposed to fuel have not been built, although some of the plutonium produced from 'civil' reactors has found its way into nuclear weapons. Plutonium stocks, of which there is much too much anyway, are expanding at a considerable rate. Many, including Bill Clinton, have warned that terrorists could use the plutonium to make nuclear bombs. It may be possible to burn up the plutonium in existing reactors, but given that plutonium fuel may be more difficult to handle if used in ordinary reactors, this may put up nuclear costs still further.

In several cases, including countries like India, Pakistan, Iraq and North Korea, civil nuclear programmes have been used as a fairly transparent cover for nuclear weapons production.

Because of these problems greater attention has been given in recent years to storing the spent fuel rods in 'dry' conditions indefinitely (perhaps at the power stations themselves) until a safe method of disposal can be found. Reprocessing and the resulting production of plutonium is thus avoided and the dry storage of used fuel rods is, in any case, rather cheaper than reprocessing. In 1993 the issue of nuclear weapons proliferation drove the Pentagon to oppose the opening of a second reprocessing plant in Sellafield in Cumbria (known as THORP, thermal oxide reprocessing plant).

THORP seems destined to be kept alive by large government subsidies. It seems likely that neither THORP nor the subsidy would be required if spent nuclear fuel rods were put into dry storage. Apart from Sellafield, the only other existing Western, civil, reprocessing plant is at Le Hague in France.

A further source of nuclear waste is the plant itself. When nuclear power stations retire they ought, in theory, to be decommissioned. The issue of what to do with the abandoned radioactive hulks that nuclear power stations eventually become has not been settled. It may never be.

Nuclear Fusion

So far, attempts to replicate the fusion processes of the sun on Earth have proved fruitless. Every few years 'breakthroughs' are announced and predictions made about the abundant availability of cheap, clean fusion power in 30 years or so.

President Peron was among the first to announce a fusion breakthrough in 1951. Fleischmann and Pons were more academically credible but no less fallible prophets of the new age (in cold form) in 1989. In 1993 scientists at the Tokamac Fusion Test Reactor at Princeton, New Jersey, announced another breakthrough. They produced 3 MW of power, although 24 MW of power was required to do this and the electricity was produced for only a fraction of a second. The project had cost $1.4 billion.[17]

Whether using eight units of electricity for every unit of electricity produced at a very high cost represents any sort of breakthrough is a matter of debate. However, it does annoy researchers into various types of renewable energy who are refused funds because their projects, which produce real quantities of electricity at measurable prices, are judged to be too expensive or too speculative.

There is controversy about the alleged cleanliness of fusion power. Although there will be no spent fuel rods to deal with, the reactors, which work with highly radioactive tritium (an isotope of hydrogen), will become very radioactive. There will be large quantities of radioactive waste and there will be the problem of decommissioning. Commercially available fusion power seems as far away now as it has been any time in the last 40 years. If it ever does arrive it might be even more expensive than nuclear fission and it is likely to result in boringly familiar arguments about how to deal with the radioactive waste that results.

For the time being, then, we cannot really sketch in a future commercial role for fusion power.

The Future for Nuclear Power

Currently, nuclear power station building plans are not taking place outside countries that have severe domestic energy shortages, or, in the case of Russia, that want to use domestic oil and gas production for export. If past records are anything to go by, government projections for nuclear power station construction will not be fully implemented.

Unless nuclear power can cut its costs it is consigned to being an energy resource of last resort. The possibilities for cost reductions depend on the development of simpler (maybe smaller) reactor designs, none of which is emerging at the moment. On top of this the political problem of nuclear waste is likely to prove to be a continuing headache.

However, there are many who suggest that nuclear power, warts and all, is better than the consequences of global warming.

The reality is that most countries may opt for carbon dioxide reduction targets that do not require widespread investment in projects with such high costs as nuclear power stations. As has been suggested earlier, it is quite conceivable that Western countries may achieve significant and continuing absolute reductions in carbon dioxide emissions purely on the basis of energy efficiency measures that do not increase consumer costs.

Some analysts insist that even extremely radical medium-term targets for carbon dioxide emission abatement can be achieved by energy efficiency measures that are cheaper than nuclear power.[18]

Although the global warming issue is unlikely to 'save' nuclear power, it may slow its decline by neutralising efforts to have existing nuclear power plants closed down (barring further serious accidents) before the end of their commercial lives. This is illustrated by the case of Sweden.

In 1980 Sweden decided to phase-out its nuclear power stations by the year 2010. At the end of the 1980s it also adopted a target of stabilising carbon dioxide emissions. Half of Sweden's electricity comes from nuclear power and half comes from hydroelectricity, so it is very difficult to phase out nuclear power and not increase carbon dioxide emissions from the electricity sector, unless renewable energy supplies can be radically increased and electricity consumption radically reduced. Moreover, since the avoidable costs of the existing nuclear power stations are relatively low, any effort to replace them with new capacity will, with the exception of the most inexpensive energy efficiency measures, involve increased costs. Environmental, economic and political factors have led to the indefinite postponement of Swedish nuclear phase-out plans.

Environmentalists can justifiably claim that energy efficiency is a much cheaper option than building new nuclear power stations. Nevertheless, it is still the case that those who call for a radical reduction in global carbon dioxide emissions have a credibility problem if their only stratagem is energy efficiency. You have to have energy from somewhere in the first place in order to use it efficiently. Can renewable energy plug this apparent gap?

12

Renewable Energy

Renewable energy has been defined as 'flows of energy occurring naturally and repeatedly in the environment'.[1]

Fossil fuel resources are limited, but renewable energy sources like solar power, wind power, biomass and water power will last as long as the Earth. Some fuels referred to as renewable are less renewable than others: some biomass use is plain deforestation; geothermal resources (the heat below ground) are extremely large, but nevertheless finite in particular locations.

Hydroelectricity aside, renewable energy seemed to be on its way out earlier this century. Nowadays most developed countries have renewable energy programmes. Renewable energy is supported because it is seen as part of the answer to many of the central problems covered by this book. It provides energy security and it can usually help combat air pollution problems. Many environmentalists see it as a non-fossil alternative to nuclear power.

Wind Power

Wind power was a popular energy source in the post-renaissance period, and has been used in places like the US West to pump water and supply power to isolated localities.

Interest in the idea that wind power could play a part in supplying modern energy requirements grew only in the wake of the 1973 oil crisis.

In their modern form, wind turbines supply electricity. Because wind speed (and thus wind energy) increases with height the turbines are placed on towers. Thus wind power is cheapest on high wind speed sites, often found near coasts and on hills. Wind power tends to be less economic on flat, inland, sites.

'Stand alone' wind turbine systems will require battery storage, and with larger systems, diesel engines to guarantee continuous power. Although this can suit some places that are not connected to the grid, wind power is, in general, most effectively deployed when the wind turbines can feed power into the electricity grid.[2]

The highest proportion of electricity supplied by wind power in any country in 1993 was 2 per cent, achieved in Denmark. The majority of wind turbines are in the US. The largest concentration of machines is in Kern County, north of Los Angeles, California, where there are 7,100 machines.

Wind power's variable output can be absorbed by the grid. Maybe up to 30 or more per cent of the network's power could be produced from wind-power without back-up or storage systems. In terms of potential technical energy output, wind power could easily supply the whole of the electricity requirements of countries like the US and the UK. Estimates suggest that use of only some of the best sites (8 m/s or above) in the US states of North and South Dakota would provide 80 per cent of US electricity requirements.[3] Studies commissioned by the European Community suggest that wind power could supply three times current UK electricity demands.[4]

The Development of Wind Power

Wind power was launched in its modern form by the so-called 'wind rush' that occurred in the early 1980s in California. 'Wind prospectors' from many countries rushed to take advantage of the tax incentives and premium price contracts offered to wind power. Wind power was then, relative to now, in a very early stage. Many of the designs were not up to scratch and many of the models did not work for much of the time. The tax concessions and generous contracts ended in 1985 and 1986, and, as the oil price fell, so did interest in wind power. Nevertheless, the wind-power industry had been established on a firm technological base upon which reliable, cheaper machines could be produced. For example, the Danish company Vestas/DWT, the world's largest manufacturer of wind turbines, now claims a machine reliability of over 98 per cent.

Interest in wind power had been growing in Denmark during the 1970s. There wind power was seen as part of the non-nuclear energy response to the oil crisis. Almost half of the wind turbines put up in California were built in Denmark.

The Danish wind-power programme really took off in the late 1980s. The majority of the total wind-power output in Denmark comes from small projects (as opposed to big wind farms in California) financed by local investors. At first grants were given. Now there is a system whereby the owners of the wind turbines receive 85 per cent of the domestic price of electricity for each kilowatt hour of electricity produced.

The number of schemes being deployed dropped in 1992 and 1993 in the face of planning problems and a decline in political interest in overcoming them. The falling costs of oil and coal imported into Denmark may have had a big effect in sapping the pressure for wind-power deployment. Nevertheless, government studies suggest that support for wind power remains high among the Danes.

The 1990 to 1993 period saw a growth in wind power in the UK and, especially, Germany. The UK has far better wind sites, but 'green' political sympathy is greater in Germany. In the UK wind-power projects are supported by being allowed to sell the electricity they produce at premium prices. The money to pay these prices comes from a small part of the 'fossil fuel levy'. This is a 10 per cent surcharge on all electricity consumers which is mostly used to subsidise nuclear power. In Germany both subsidies and high prices are given.

The Netherlands has also developed a wind-power programme, although finding sites in this densely populated area has often been difficult. Indeed, planning problems, centring on noise and especially on visual impact, have slowed, although not halted wind power's progress in thickly populated Europe. In the UK there have been fierce arguments over many of the planning applications for wind farms in windy areas like Cornwall and Wales.

Although the costs of wind power had greatly declined by the early 1990s compared with the early 1980s, low natural gas prices meant that even in the most favourable circumstances commercial prices for electricity from wind power were still much more expensive than electricity from combined cycle gas turbines (CCGTs).

The amount of electricity produced by wind power is still small relative to fossil and nuclear fuels. By the end of 1993 there was roughly 3,000 MW of wind power worldwide (see Figure 12.1). Because the wind is intermittent turbines have a lower capacity factor than conventional power stations. Thus the whole of the world's wind power in place at the end of 1993 produced little more electricity than that produced by a single large nuclear power station.

Clearly, wind power has a long way to go before it is transformed from being an interesting (and to some, irritating) curiosity into a major force in the energy world. Whether this happens depends on whether wind power can be both cheap and environmentally acceptable.

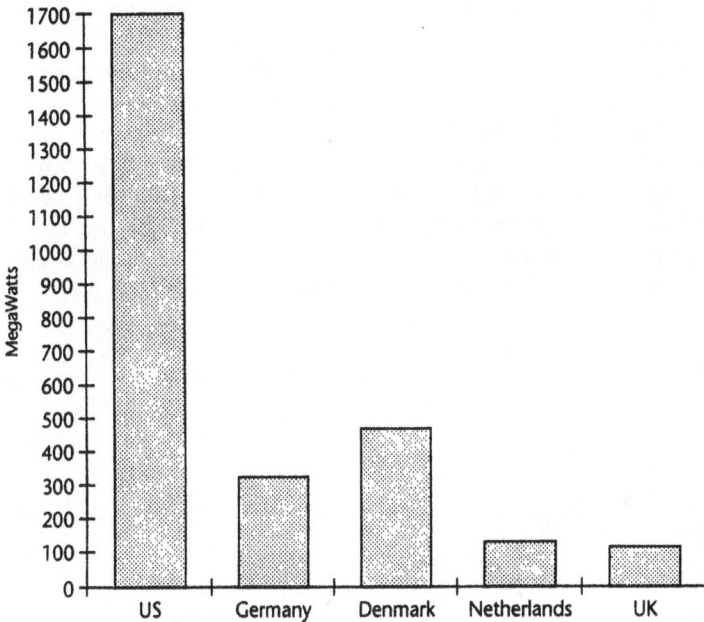

Figure 12.1 Leading Wind-power Nations
(Installed Capacity end 1993)

Source: Various issues of *Wind Power Monthly*

The Costs of Wind Power

The fuel for wind power costs nothing and the running costs for the largest new schemes are around 1 c/kWh in the US. As is the case with nuclear power, wind power is very capital-intensive, although wind turbines can be made and deployed within a few months.

The costs of wind power have been coming down (as can be seen in Figure 12.2), because the efficiency of wind turbines has increased and the cost of making them has fallen. Because of the intermittent nature of wind, bankers tend to assess wind-power projects using higher discount rates than, say, CCGT projects. For example, in the UK discount rates of at least 12 per cent and maybe 9 per cent in the US would be applicable even for large (say, 50 MW) schemes.

The lowest bids for wind-power contracts made in the UK government's 1994 round of renewable energy contracts were for around 4 p/kWh. This represented a big fall compared to the bids of at least 9p/kWh made in the previous 1991 round. The main reason for this decline, apart from the much longer duration of the contracts on offer, was the prospect of a new generation of 600 kW wind turbines. These have much lower relative capital costs than previous, smaller, turbines. In California wind-power contracts were being accepted by developers in 1993 for around 7 c/kWh, although by 1994 US wind generator manufacturers were claiming that in the case of big wind farms the costs of electricity from wind power could be reduced to below 5 c/kWh.

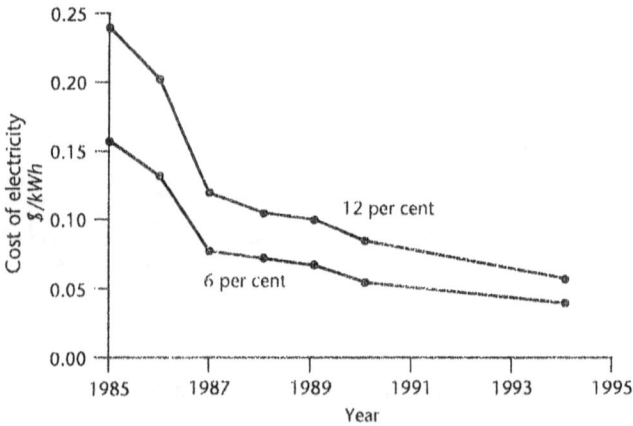

Figure 12.2 Declining Cost of Electricity from Wind Turbines in California

Note: dotted lines indicate estimates. 6 per cent and 12 per cent labels refer to discount rates. Typical wind speeds in Californian mountain passes are assumed

Source: Cavallo, Hock and Smith in Johansson et al. (eds), *Renewable Energy* (London: Earthscan, 1993), p. 122

Since the early 1980s the optimum economic size for wind turbines has risen from 50–100 kW to between 200 kW and 600 kW in output and the costs of wind power have declined consistently, as can be seen in Figure 12.2.

Apart from capital costs, the other key element in wind-power economics is the average wind speed at a particular site. A given wind turbine at one of the best sites on the North German coast (say 6 m/s annual mean wind speed at the turbine hub height) will produce no more than 60 per cent of the electricity output achieved at 8 m/s (hub height) on a windy ridge in central Wales or California.

Unless there is an appropriately sized electricity load to be served on a consistent basis, wind turbines need to be connected to the electricity grid. This can be too expensive if the wind-power project is too far away from the right sort of transmission line. Generally speaking, this will be less of a problem with larger wind farms, but even here there are limits. Larger wind farm developments will be able to achieve big discounts from manufacturers of up to 30 per cent compared with the retail price of single turbines.

It can be seen from this that the most economic form of wind power consists of large projects sited in the most windy conditions. However, in densely populated Western Europe it is difficult to gain local acceptability for very large schemes. It has been demonstrated in Denmark that local ownership of wind turbines makes it easier to gain planning permission, which may offset the fact that these smaller projects are, generally speaking, more expensive than large projects.

Environmental Impacts

The environmental controversies surrounding the impacts of wind power have mostly focused on three issues; visual impact, noise and bird deaths.

There is little wind-power developers can do about the visual impacts of wind turbines. The turbines need tall towers. A 100 kW wind turbine will usually have a tower and hub height of around 24 metres and a 500 kW wind turbine will usually sit on a 40 metre high tower.

The land area covered by wind farms is considerable. In order to produce the same electricity as a large (1,000 MW) conventional power station, a wind farm would require the use of around 500 square kilometres, although this area can still be farmed. Nevertheless it can be imagined that in densely populated regions the visual impact of the wind turbines presents political problems. In the UK the provision of around 10 per cent of electricity supplies from wind power would need around 1 per cent of the UK to be covered by wind turbines, although in the US this figure would be roughly 0.3 per cent.

Concern over the impact of wind farms has, generally speaking, been greater in densely populated areas of Europe than in the US. In Europe, countryside areas (where higher wind speeds are generally to be found) are relatively scarce and are highly prized for their recreational value. Planning constraints have been softened in Denmark by a system of locally financed wind-power cooperatives through which the majority of wind turbines have been deployed. This has meant that local people have been given an interest in the developments.

By contrast, landscape issues have, with some notable exceptions, not featured so highly in the usually sparsely populated US where planning applications for wind farms have enjoyed a high success rate.

Again, noise issues are more important when the number of people living close to the wind turbines is significant. Although wind turbines produce little noise and can be barely heard at distances over 300 metres, even slight noise can be an annoyance if someone has to live with it on a permanent basis. Wind-power supporters say that the noise from motorways and other major roads is much louder at the same distances.

The noise emitted by turbines varies with size, type and modernity. The most wind turbine noise disturbance tends to come from the gears rather than the swishing of the blades themselves. The gears produce low frequency noise which is the most troublesome. Enercon, the leading German wind turbine manu-facturer, is selling a gearless, 500 kW machine, which it claims is very quiet.

Controversies over bird deaths have been limited to a small number of sites. However, it is an issue in the US. The giant Altamont Pass site in California has been the subject of studies and several birds, chiefly birds of prey, have been classified as having been killed by the blades. There has been little or no controversy over bird deaths in Europe, with the exception of a wind farm at Tarifa in Southern Spain by the Straits of Gibraltar. Large numbers of birds travel through the Straits.

The power lines are a hazard to birds, although this is common to all power installations with overground power cables. There are measures that could be taken to limit bird deaths in sensitive locations. One idea is to paint the turbine blades to increase their visibility, although this obviously increases their visual intrusion.

One means of lessening the impact of wind turbines is to site them in shallow waters offshore. In fact the Danes, Swedes and Dutch are developing offshore wind-power demonstration projects. The Danes have established an eleven turbine 4.5 MW wind farm just off the coast near Vindeby. The electricity from this project is around 50 per cent more expensive than from comparable Danish onshore wind farms. It has been suggested that the costs of building underwater bases for the turbines could be avoided by making the machines sit on floating bases anchored to the sea bottom. The offshore wind-power resource is very considerable for countries like Denmark and the UK. In 1994 the UK government rejected the idea of funding any demonstration offshore wind farms.

It can be seen from this analysis that the environmental impacts of wind power are most likely to constrain its development in the more thickly populated regions of the world.

The Future of Wind Power

The progress of wind power seems to be proceeding in ebbs and flows, rather like a fashion which comes in and out of favour. The Danes set themselves the target of producing 10 per cent of their electricity from wind power by the year 2005, but with the slowdown in deployment it is unlikely that this target will be met. In 1994 there was a rush of applications for planning permission for wind-power schemes in the UK. This resulted in a controversy about the environmental impact of wind power, although the number of planning applications was around ten times larger than than the number of wind-power contracts that the UK government said it would award.

However, there are some more promising signs for the future. In the US a new wave of wind-power projects is, at the time of writing, taking shape. At the end of 1993 some 600 MW of wind-power bids were accepted in California. Other states, including Minnesota, New York, Oregon and Iowa, have put together firm plans for deploying wind-power schemes. US Wind Power, the wind turbine manufacturer, has assiduously lobbied Public Utilities Commissions about the advantages of wind power in providing energy supply diversity and clean energy.

Some power industry brokers in the US are being persuaded that wind power can actually bring cost advantages. Given the potentially volatile nature of natural gas prices, with new coal plant being weighed down by the cost of meeting clean air requirements, and nuclear power unlikely to be a serious competitor, many believe that renewable energy in general and wind power in particular seems worth looking at in purely naked commercial terms. US wind power is also being helped by a 1.5 c/kWh subsidy. Wind-power schemes on the best sites in the US are, with or without subsidy, little or no more expensive than new 'clean coal' power plants.

Wind power also has some hidden cost advantages because it can be deployed quickly and in (relative to conventional power stations) small packages. Often small projects sited near the end of 'weak' rural power distribution lines can strengthen the system, so giving them a cost advantage.[5]

It is because there has been a tremendous decline in wind-power costs that wind power is beginning to be taken seriously. There is every reason to expect that its costs will continue to decline and that it will become a very attractive energy resource to those countries that have good wind resources and relatively low population densities. Indeed, these criteria are fulfilled by the majority of major countries outside Western Europe, and even in these 'old world' states the allure of cheap power prices and an increasing desire to cut down on carbon dioxide emissions and energy dependence could give the wind-power industry a fresh boost. In Europe community funded and cooperative wind-power ventures may overcome many planning objections.

Nevertheless, the irony may be that developing countries might, in a few years' time, start manufacturing and deploying their own wind turbines in massive quantities because they will become cheap, indigenous energy sources, while in Western Europe wind power grows much more slowly because of environmental constraints.

There are already some straws in the wind. Argentina has announced a 500 MW wind-power programme in collaboration with Micon, the Danish wind turbine manufacturer. The Danish method of community financing of wind-power schemes is to be employed. The US-based New World Power Corporation claims it has reached agreement with Mexican landowners to begin construction of a 20,000 MW wind-power programme. This would rival even the world's largest conventional power projects in size.[6]

Solar Power

The US receives roughly 5,000 times as much energy in sunlight as it consumes in electricity in the course of a year. The energy in all the electricity consumed by the UK is only a thousandth the size of the UK's supply of solar energy from sunlight.

Of course, some places receive rather more solar energy than others. Southern US will receive between 2,000 and 2,500 kWh a year per square metre compared to 1,000 kWh in the UK. Thus solar power, in all of its forms, is usually more economic in hot parts of the world than in cooler zones.

In some ways it is odd talking about solar power being an underused resource: how do we think our crops grow or we survive?

Passive Solar Energy

The simplest way to utilise solar energy is to design buildings to make the best use of sunlight. This is called passive solar energy. Glass traps the warming infra-red radiation and retains the energy inside the building. Large windows in a building whose main façade faces the south can maximise sunlight capture. Shading should also be employed to afford summer cooling.

The impact of passive solar features is greatest when passive solar techniques are combined with energy conservation. Indeed, passive solar energy will only make up a high proportion of energy consumption when the energy efficiency of the building is maximised. This is achieved through having thick insulation, high-mass building structures to maximise heat retention, double or triple glazing and use of low emissivity glass that reduces heat being radiated from the inside panes of glass to the outside.

Another energy-saving aspect of passive solar is daylighting. This can cut down on electricity requirements for lighting rooms.

All these techniques need to be carefully balanced with each other, and orchestrated to fit in with the local climate. This will happen only with carefully organised building regulations and architects who are well steeped in the techniques of solar, energy-efficient buildings.

Active Solar Energy

Active solar energy involves capturing the solar energy with a specific collecting device. In the case of buildings this usually means flat plate water collectors fixed on roofs or other surfaces. The solar energy is absorbed by a material made of a black substance and carried away by a liquid. Sometimes alcohol is used. It has a low boiling point and carries the heat to a heat exchanger where the energy is transferred to water.

The usual, and most economic, use for active solar heating is to provide hot water in warm climates. Most houses in Cyprus and a high proportion of houses in Israel and Greece have solar water heaters. Building regulations in Israel have been amended to make it mandatory for all buildings less than ten storeys high.[7]

Active solar systems can be economically used for space heating only in continental areas that have cold but sunny winters.

Solar water heating took off in warmer parts of the US in the early 1980s. This revived an industry that had flourished before the widespread availability of cheap oil and natural gas for space heating in the 1920s and 1930s. However, after oil prices fell in 1985 and incentives for renewable energy were removed, the market deflated once more.

Active solar systems are fully competitive with fossil fuel (and especially electrical heating) systems in warm countries where fossil fuels are relatively expensive. Their use could become more widespread in these circumstances if consumers' pay-back times were shortened with the help of subsidies or mortgage tax incentives. The theory behind this is exactly the same as the theory which supports subsidies being given to energy efficiency methods. The subsidies or rebates can even out the difference between pay-back times used by energy suppliers and energy consumers.

Such policies are employed in relatively cool Netherlands. There, energy companies are doing a good trade in solar water heaters. Forty per cent grants for the heaters plus a high level of green consciousness among many consumers are the reasons for this phenomenon.[8]

In the UK active solar is not supported by incentives, the absence of a reasonably sized market prevents manufacturers from cutting unit costs and the climatic conditions mean that active solar systems are economic only if they are installed on a DIY basis.

Solar-thermal Concentrating Systems

Usually active solar systems are limited to low-temperature applications. Yet many attempts have been made to develop solar systems that concentrate solar energy to achieve higher temperature heating. These are usually referred to as solar-thermal concentrating systems.

It is said that Archimedes organised Greek soldiers to use their shields to concentrate sunlight on, and set fire to, Roman ships attacking the fortress of Syracuse in 212 BC. More modern experiments have confirmed that this is possible.[9]

Large quantities of high temperature heat can be derived from solar concentrating systems in sunny climates. The most widely applied concentrating technique is the parabolic trough which involves a trough-shaped mirror/reflector. The sunlight is focused on a tube carrying liquid along the centre of the parabolic mirrors. This can be used to provide heat for industrial purposes, and is used today by some public institutions (like prisons) in the US which can afford to invest in projects with relatively long pay-backs.[10] Such systems were competitive with fossil fuels during the early 1980s.

The biggest use of solar-thermal concentrating systems has been in generating electricity. High temperature heat turns water into steam which is then used to drive a turbine to generate electricity.[11]

During the 1980s the state of California gave some generous tax concessions to allow the deployment of several solar-thermal electric power units. A company called Luz put in place several parabolic trough projects. The largest scheme has a generating capacity of 80 MW. The projects involve using the concentrated solar energy to raise steam to drive turbines. Natural gas is also burned to keep the plant running when there is no sunshine. The costs of the electricity from the final project came down to 9 c/kWh compared to around 27 c/kWh for the first project.[12]

However, natural gas prices fell, state tax credits ended and the company went bust in 1991. The company left 354 MW of solar power plant all of which are, at the time of writing, still operating today in the Mojave Desert. They produce most power precisely at the times when the electricity companies most need it to feed the voracious appetites of air-conditioning systems. These projects take up a lot of land. An 80 MW project takes up nearly a square mile, although this is not a big problem in desert areas where the projects have been sited.

Before the tax credits ran out, Luz was hoping to reduce costs further by building a bigger 200 MW project. This would have used a combined cycle plant (to burn natural gas directly). It was also planned to use the concentrated solar energy to heat steam directly rather than employing an intermediate heat-carrying mineral fluid.

There are other solar-thermal concentrating power plants of slightly different designs such as solar dishes (which are shaped a little like radar dishes) and central receiver systems where numerous dishes focus solar energy on a central tower.

The trouble with solar-thermal concentrating systems is not that they can never improve their efficiencies (currently around 12 per cent of the solar energy is turned into electricity) or that they cannot reduce their costs; the chief problem is one of markets. Wind turbines and other forms of solar power have their own niche markets or, alternatively, can be put into practice in relatively small packages. Wind-power costs are, in any case, becoming moderately competitive in prime conditions. But improving the economics of the solar-thermal concentrating systems depends on building larger projects since, among other things, the efficiencies of steam turbines and combined cycle systems improve with the size of the project. Thus the chances of seeing such technology deployed at its optimum size is constrained by the relatively modest sizes of the funds available for renewable energy development programmes. You can have lots of wind turbines and photovoltaic solar modules with $50 million, but you will not get a solar-thermal electricity system built at the right size.

It has been suggested that solar-thermal concentrating systems could be built using dishes and small Stirling engines. Stirling engines are an old, but little used, existing engine design that involves external combustion. This means that the moving parts of the engine are contained inside the machine. However, Stirling engines are still very expensive and difficult to maintain. Advances in material technology are needed before they can become cheap enough to supply power to the grid. Nevertheless, at a projected price of 15 to 20 c/kWh[13] such systems may one day be relevant to the provision of electricity to areas not connected to the electricity grid.

Solar Photovoltaics

Photovoltaics, or PV, for short, are the newest and most 'high tech' forms of solar power. PV involves the direct conversion of sunlight into electricity using solar cells. PV has a much more glamorous image than other solar energy systems, despite the fact that electricity from PV systems is still more expensive than electricity from, say, Luz-type solar-thermal concentrating systems.

The PV effect was first documented by Becquerel in 1839 and in the 1880s selenium cells were designed which were used later in photographic exposure meters. Modern PV results from research into semi-conductors (which are used in transistors) in the 1950s. The first PV cells were used in the space programme, but since the 1960s PV have gradually penetrated a number of markets in communications, pocket calculators, various other consumer goods and off-grid electricity supplies. Silicon is the most favoured material for PV systems.

Silicon is a semi-conductor because while it conducts electricity (through a flow of electrons) better than wood, it is not as good a conductor as, say, copper.

Essentially, PV systems work through photons of light striking and displacing electrons, leaving a hole in the cell's electron structure. An electric charge is created, although it will be quickly neutralised unless the hole and electron are drawn apart and separated. Semiconductor junctions achieve this separation. This creates a barrier across which electrons and holes pass moving in opposite directions, thus creating a current.

Some systems use several layers to create several junctions. Using more than one junction increases the number of wavelengths of light utilised. Hence the conversion efficiency increases. These systems involve so-called 'multijunction' cells.

PV Technologies
There are now several types of competing PV techniques. Monocrystalline silicon cells have, until recently, been the usual PV technology. This involves carefully grown single crystal cells. The crystals are sliced and incorporated in flat plate cells. The highest level of efficiency of conversion of solar energy into electricity is around 17 per cent in (1993) commercial operation, although rather higher efficiencies have been achieved in the laboratory.[14] However, monocrystalline PV is also the most expensive.

Growing many small silicon crystals at the same time is a much cheaper process. This produces polycrystalline cells. However, these have so far achieved efficiencies of only around 10 per cent in commercial practice. Polycrystalline silicon is often used as material for so-called thin films.

There is increasing interest in so-called thin film PV. It is potentially cheap because it can be laid on material such as glass and can easily be matched with the outside of buildings. Thin films can be made out of several materials besides polycrystalline silicon: gallium arsenide, copper-indium-diselenide and cadmium telluride are all possible candidates. Gallium arsenide is very expensive but the theoretical potential efficiencies of these cells is very high. Experimental solar cars have performed well using this material. Thin film PV can also be produced using so-called amorphous silicon.

Amorphous silicon, that is silicon in non-crystalline form, has emerged as a major player in the PV world. Amorphous silicon is cheap to produce relative to other materials. Its drawback has been its low efficiency and the tendency of efficiencies to degrade over time. Nevertheless, the most recent designs can achieve efficiencies of up to 9 per cent on a consistent basis when alloyed with other materials. Amorphous silicon is already heavily used in applications such as calculators where low cost is important and efficiency does not particularly matter.

PV technologies are now relatively commonplace. PV equipment is now a standard measure for powering telephones and cooling medical supplies in remote areas, pumping water, buoys, consumer products such as wrist watches and calculators and providing electricity for buildings which are not connected to the electricity grid. Around 50 MW of PV were being manufactured every year at the beginning of the 1990s and this proportion is expanding quickly. How has this capacity been used?

Deployment of PV Solar Power
PV schemes have been deployed in sizes ranging from a few watts for telephones in remote areas to 'central schemes' of over 1 MW in size. The US, Italy, Spain, Japan, Germany and Switzerland have funded a range of demonstration grid connected schemes, some of which have involved use of concentrator systems. This allows relatively low cost concentrators (dishes or heliostats) to collect sunlight and focus the sunlight on to some high efficiency flat plate collectors.

National and multinational overseas aid agencies and the World Bank have funded a number of schemes to provide electricity to off-grid communities and facilities in the developing world. It is certainly true that connecting remote rural areas to the grid can be extremely costly, especially if the electricity demand is very low.

Unfortunately, some projects sited in developing countries (subsidised by Western aid) have been poorly planned, with little attention paid to how the projects are to be maintained. Very often the schemes are much more costly than diesel generators, which is the main means of supplying electricity away from the grid. At the end of the 1980s a 10 kW PV system was likely, at a 10 per cent discount rate, to supply electricity at around six times the cost of supplying power from an equivalent-sized diesel generator.[15]

Although the costs of PV are declining,[16] it is likely that PV will spread more quickly if it concentrates on the 'niche' markets for which it is currently suited rather than being discredited in the form of inappropriate schemes forced on poor villages by the West.

There are a growing number of markets for PV in off-grid situations where only very small amounts of power are needed and there are other markets in places that are difficult to supply regularly with diesel. PV is at least 50 times cheaper than dry-cell storage batteries.[17] It is thus cost-effective in telecommunications and other information systems, refrigeration, small-scale pumping and also in supplying power for consumer goods such as televisions and radios and lighting in remote areas.

PV seems also to be making progress in places like the remote islands that make up the Pacific nation of Tuvalu. By 1993 around a quarter of Tuvalu residents were supplied by PV, and the programme is being rapidly extended.[18]

Manufacturing PV cells is still a very expensive process compared even to making wind turbines. The PV manufacturers which currently exist have only short production lines and high overheads. Current assessments put the cost of PV in ideal circumstances in the sunniest locations of the Southern US as being at least 25 c/kWh, and these costs do not include the cost of storage systems that are necessary for off-grid systems. In the UK the cost would be over 40 p/kWh.

A sort of halfway house between supplying power to the grid and the stand-alone system might arise when PV can be used to help the consumer offset the retail costs of electricity and export any excess to the grid. The cheapest way of doing this may be to incorporate the PV systems into the building structures themselves.

PV on Walls and Roofs

The amount of power that could be generated from fitting PV to roof and building cladding could constitute a major proportion of total electricity demand even in cold countries like the UK. One experimental project in the UK involves 40 kW of PV panels being fitted to the façade of the University of Northumbria's central computer building in Newcastle upon Tyne to replace slabs of concrete that were falling off.[19]

In hot, sunny countries where the peak demands (which are expensive to supply) are in the daytime because of power-hungry air-conditioners, PV will be very much suited to electricity demand profiles.

United States Solar Systems has tested a relatively high efficiency 'multijunction' amorphous silicon alloy-based system which it believes will deliver efficiencies of 9 per cent on a long-term basis. Thin films of silicon are laid on steel strips, as can be seen in Figure 12.3. The steel can be mounted as rooftiles, façades or set up on the ground. The development of relatively high efficiency amorphous silicon cells is important since amorphous silicon can be manufactured much more cheaply than other materials used in PV technology.

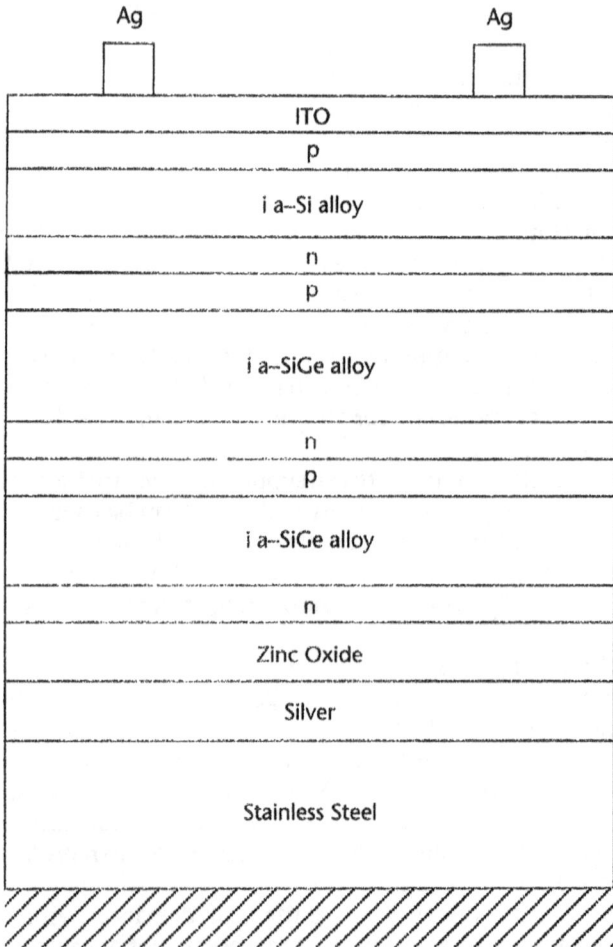

Figure 12.3 Schematic Diagram of Triple Cell Structure of USS Amorphous Silicon Thin Film Alloy System

Note: there are three cells, each consisting of three layers which are less than one micron thick, laid on 5 mm thick steel strips

Source: United Solar Systems Corp., Troy, Michigan, 1994

The Future of Solar Photovoltaics

The US Department of Energy has estimated that when company production lines grow to 10 MW a year, PV costs should decline to around 16 c/kWh for electricity supplied from large arrays in locations such as New Mexico in Southern US. This assumes a 9 per cent discount rate and 10 per cent cell efficiency. It was expected that a 10 MW facility in Newport News, Virginia, would be operational in 1995.[20]

Meanwhile Australian researchers at the University of New South Wales have claimed that the costs of solar electricity could be cut by 80 per cent within a decade using new techniques that allow conducting metal to be buried inside rather than on the surface of the solar cells. This will allow more sunlight to reach the silicon. Thin film crystalline silicon cells could, using these techniques, produce power with (improved) efficiencies of over 15 per cent at a cost which is five times lower than that of conventional crystalline silicon cells.[21]

Declining costs are likely to open up new markets in several areas. The first is expansion of the already existing market for stand-alone PV systems for off-grid situations where PV will compete more effectively with rural electrification programmes and diesel generators. Of course, 'off-grid' situations can exist in trains or in the back seats of cars as users of cellular telephones and laptop computers struggle to maximise the amounts of time they can squeeze out of their energy-short and expensive nickel-cadmium batteries! Samsonite Corporation and Sun Wize Energy Systems, for example, have come together to produce a small, thin, lightweight solar panel attached to a briefcase. Such equipment would use sunlight or artificial light to provide power for computers and cellular phones.

In many ways solar PV's advance is a chicken-and-egg problem. Its costs need to come down to capture a big market, but it needs a big market (and all the advantages that bigger production lines bring) to bring its costs down rapidly.

However, markets for solar PV are growing. If the costs decline to the price of domestic power (say 12 c/kWh), grid-connected consumers in sunny places can start using the systems to cut their own power bills, perhaps with assistance from agencies offering long-term lending facilities.

In a few years' time we shall see costs come down to allow dwellings in new housing estates (initially in areas like New

Mexico) to be built with PV panels and inverter systems so that households can send excess power to the grid.

Biofuels

Biomass is produced through photosynthesis. Sunlight is used by the biological 'machine' chlorophyll to combine carbon dioxide with water to produce oxygen and carbohydrates. Carbohydrates are the building blocks of plant matter.

Biomass material is discarded as waste products by both animals (dung) and by industrial society (rubbish).

Biomass is the oldest fuel of all. Indeed it still accounts for at least 15 per cent of global energy use and about two-fifths of energy use in developing countries.[22]

In the industrialised world biomass has long been overshadowed by fossil fuels, but there are some promising and intriguing possibilities for its return. Biofuels are versatile and can be used to supply heating and electricity. They can be used in solid, liquid or gaseous forms.

Environmental Issues

If biomass is harvested renewably (that is, any stocks harvested are replaced with seedlings that will replenish the stock), then no net carbon dioxide additions to the atmosphere will be made by burning it.

If wood or other organic material is left lying around in the open air it will biodegrade, giving off carbon dioxide. If rubbish is buried in landfill it will generate methane which is a powerful greenhouse gas. Thus use of biomass waste as an energy source makes no net addition to so-called greenhouse gas emissions.[23]

Biomass usually contains only very small amounts of sulphur compared to coal. When burned it will give off significant NOx and dust emissions, although less than in the case of coal. Like coal, such emissions can be scrubbed. Biomass sources are dirtier than natural gas, which produces no dust emissions, although renewable biomass energy sources result in much lower carbon dioxide emissions than natural gas.

Since biofuels and biofuel technologies differ widely it would be hazardous to generalise further about environmental issues, and I shall deal with specific environmental concerns as they relate to the particular type and use of biofuels.

Biomass Residues

Heavily forested countries like Sweden and Austria derive roughly a tenth of their energy needs from residues resulting from forestry activities. The US provides around 4 per cent of its energy requirements from biofuels. The UK's biofuel use is small compared with total energy consumption.[24]

Many studies suggest that energy supply from biofuels could be significantly increased in all countries. The productivity of existing US forestry practices could be massively improved by removing unharvested low quality stock. This could yield wood with an energy value of between 12 and 28 EJ each year. Total US energy consumption is around 85 EJ a year.[25]

There are many unused energy resources in the form of vegetable, animal, household and industrial wastes. Denmark already provides 1.5 per cent of its national energy needs from straw burned in straw-fired boilers. These boilers are often connected to district heating schemes [26]

In Sweden and Switzerland over 80 per cent of unrecycled household waste is burned to provide energy, while at the other end of the scale the UK used only 2 per cent of household waste in waste-to-energy schemes in 1992. Burning waste is a controversial topic, partly because of fears about emissions of toxic substances like dioxins and furans, and partly because many would prefer to see garbage recycled rather than burned for ideological reasons. There is a debate about the amount of waste that can be recycled without using more energy to recycle the waste than is saved by recycling the material.

Advocates of waste-to-energy schemes point out that emissions regulations in the US and Europe have been tightened very greatly in recent years. Around half the costs of new energy-to-waste schemes in the US are tied up with emission control systems.

Many say that the bulk of waste that cannot be economically recycled should be burned with energy recovery rather than consigned to landfill sites that can cause methane leaks (these cause global warming and can explode) and acidic leachate fluids that result from their anaerobic decomposition.

There are massive wastes resulting from the forestry and sugarcane industries in developing countries that could be utilised, especially if biomass combustion techniques can be improved (this issue will be discussed later).

Energy Crops

The whole of the world's commercial energy requirements could, in theory, be supplied by planting around 9 per cent of the planet's surface area with fast-growing trees. This assumes 12.5 dry tonnes/hectare of annual wood production at 20 GJ/tonne. Fast-growing trees produce much more wood than trees grown by conventional means, but fast-growing tree plantations can be expensive to cultivate and manage. So-called coppicing is a favourite method of growing trees quickly. The trees are planted once every 20 to 30 years and the trees are harvested every three to five years.

Energy crops take up very large quantities of space, much larger even than wind power or solar power to fuel the same energy output. This is because the photosynthetic efficiency of energy crops is very low. Even in tropical areas with all-year growing seasons the proportion of sunlight falling on a given area which will be turned into plant matter will not exceed 2 per cent.

Environmentalists sometimes criticise energy cropping because, as in other forms of agriculture, fertilisers and herbicides are used.

Besides fast-growing trees like poplar, willow and eucalyptus, crops such as sorghum, elephant grass and sugarcane can be used to provide energy.

Studies suggest that in temperate regions of the world delivered fuel from fast-growing trees would cost £2.65 to £3.08 per GJ (6 to 10 per cent discount rate),[27] which is too expensive to compete with fossil fuels, especially given the latter's lower transport costs.

On the other hand, studies carried out in the Pacific Northwest of the US, using a rotation period of eight to ten years and hybrid poplar clones, suggest that production could be increased[28] from the usual 12 tonnes or so per hectare per year to around 20 tonnes per hectare on a fully commercial basis. This could, if successfully commercialised, bring costs down to under $3 per GJ.

In Sweden the planting of fast-growing trees is being subsidised (on, as yet, relatively small areas of land) to provide fuel for the large existing wood fuel market. In the UK there are efforts to grow coppiced willow to fuel small power stations funded under the UK's renewable energy programme.

Energy crops are going to need subsidies and access to markets if they are going to be commercially viable in the developed world. However, in warmer, developing countries, higher yields can make energy cropping more attractive, especially if the country has a large energy import bill.

There are some extensive fast-growing tree plantations in Ethiopia, used to supply Addis Ababa with fuel. Fast-growing trees are usually used to provide feedstock for the pulp, timber and plywood industries. In Brazil they are used to make charcoal. There are around 800,000 hectares of fast-growing eucalyptus plantations in Spain and Portugal which supply the pulp industry.[29] Spain and Portugal are dry countries and there are complaints that the trees soak up too much water.

Perhaps the best known biofuel programme in the world is the Brazilian pro-alcohol programme. Sugarcane is grown and fermented to make alcohol which acts as a substitute for gasoline. Seventy dry tonnes/hectare per year are routinely produced. This productivity is several times higher than is achieved in temperate regions with fast-growing trees.[30] In tropical zones there is more solar radiation and there is a longer growing season, in addition to which sugarcane is an extremely productive crop in sunny countries. There is a big market for fuels which can alleviate Brazil's dependence on oil imports.

Clearly there is a massive potential for energy crops, but such resources will remain largely unused unless there are markets and, in many cases, subsidies to enable the fuels to compete with conventional alternatives. Some analysts, such as those connected with the LEBEN (Large European Biomass Energy Network), believe that in places like Europe non-food biomass plantations will in future be cultivated in order to produce a mix of industrial products. Biofuels may be supplied as one of the by-products rather than the main product itself. The increasing amount of surplus agricultural land in the developed world presents an ideal opportunity to transfer subsidies from food towards these activities, thus keeping farmers in business.

Direct Burning

The type and quality of delivered energy supplied by biofuels is crucially dependent on the biomass technology employed. The traditional way of producing energy from biomass is through direct burning, but the bulk of biofuels used today are burned in very wasteful and smoky ways by poor people in the developing world. Cleaner, more efficient wood-burning stoves have been successfully introduced in places where the prices of woodfuel are high (such as Rwanda), but there has been much less interest in using these stoves in areas where woodfuel is readily available at little or no cost.[31]

Modern wood-burning stoves are used in developed countries, especially Sweden and Austria. Here wood is a major fuel for domestic heating and cooking, district heating schemes and industry. There are some wood-fired cogeneration systems in Sweden, and the Swedish government gives grants to support such projects.

In Sweden there are strict controls on the dust emissions that can be a problem with biomass burning. Large or medium-sized installations can be effectively regulated. Unfortunately it is more difficult to ensure that domestic boilers and stoves are regularly maintained to stop their cyclone dust collectors becoming clogged up and thus ineffective.

There is also considerable domestic use of wood in many areas of the US, but this is declining as natural gas pipelines extend.

There has been a recent growth in the amount of heat and power generated from small power units sited in the timber and pulp plants in the US and the pulp and sugarcane industries in many developing countries. These mini-power stations burn the wood waste (or bagasse, fibrous waste left over from sugarcane processing) and produce useful amounts of electricity for on-site use. In the US some electricity is sold via the electricity grid under PURPA rules. More often than not biomass is burned as part of a cogeneration plant. There are over 8,000 MWe of biomass power stations in the US providing roughly 1.5 per cent of US electricity. Three-quarters of the plants are based in the timber and pulp industry and the rest burn domestic waste.

The relatively small plant used in these operations are inefficient producers of electricity (invariably less than 20 per cent of the energy in the fuel input will be turned into electricity) but low efficiency does not matter so much if the fuel source is very cheap. Burning the biomass often avoids significant disposal costs.

More efficient steam-raising wood-burning power stations are very expensive at the small sizes relevant to most of the processing factories and it is impractical to have very large (say over 100 MW) wood-fired power plants because of the limited availability of wood in a given area. Wood has a much lower energy density than coal and costs many times as much to transport. The problems of fuel availability and transport costs are crucial difficulties facing efforts to expand biofuel use.

Designs have been proposed for 'whole tree burners' that could increase efficiency to 36 per cent, but these may not be very competitive below a 60 MW size.[32]

Of course, wood can be used with a good degree of efficiency to provide heat-only energy, or, in cogeneration systems, to produce both electricity and heat.

There are various ways in which biomass material can be processed, through heating, so as to convert the biomass into different forms of fuel. Pyrolysis has traditionally been used as the method of producing charcoal. In a more modern form pyrolysis can be used to provide a range of liquid, gaseous and solid forms of biofuel. Liquefaction is a means of turning the biomass material into liquid form under high pressures.

Gasification

In recent years, gasification has attracted most attention as a means of utilising energy from biomass. Coal gasification involves passing steam over the coal to produce a low heating value synthetic gas (called town gas) consisting of methane and carbon monoxide. Town gas has been produced in this way since the end of the eighteenth century.

Town gas has been mostly replaced by natural gas. Nevertheless, the gasification process can be used to turn biomass into a synthetic gas. This gas can be used as a fuel itself or it can be further treated so as to produce a variety of products including methanol, alcohol and ammonia.

The costs of producing electricity from biomass may be reduced if biomass gasification is linked to gas turbine technology. The marriage of coal gasification and gas turbines in the form of IGCC plant has already been described in Chapter 9.

Biomass is, theoretically, easier to gasify than coal and, unlike coal, it does not require expensive processes to remove sulphur. Biomass-integrated gasification/gas turbines (BIG/GT) should, in theory, be cheaper than coal-fired equivalents. Moreover, BIG/GT plant could, in theory, be built at relatively modest sizes for relatively low costs and still deliver impressive levels of energy efficiency. Although a great deal of effort has gone into commercialising IGCC plant using coal, less attention has been given to developing BIG/GT plant. Nevertheless, a number of demonstration projects are being put into practice in places like Finland, Sweden, Brazil and the US.

General Electric, for example, projects that a typical BIG/GT plant would come in the 15 to 50 MW size and would generate electricity at an efficiency of 34 to 38 per cent. General Electric has been experimenting with wood and bagasse as fuels for BIG/GT plant.

They estimate that around 10,900 hectares (109 km^2) of fast-growing trees would be required to fuel a 30 MW plant.[33] BIG/GT technology can be adapted for combined cycle and cogeneration modes, which would improve overall efficiencies still further.

One estimate suggests that if all of the residues from the sugarcane industries in 80 countries were used to produce electricity in advanced BIG/GT plant, then such plant would produce over a third of the electricity consumed by these countries in 1987. If BIG/GT plant could be built for the same price as (coal) IGCC plant, then electricity could be produced for about 5 c/kWh, an attractive price for countries with large energy import bills.[34]

Electricity is already being produced from small (under 1 MW) gasification plant using 'off the shelf' internal combustion engines. They convert a higher proportion of biomass energy into electricity compared with small steam-fired plant and are thus likely to be more economic in many circumstances.

As is the case with coal-fired IGCC plant, toxic tar products will be left behind by the gasification process if so-called 'updraft' gasifiers are used. 'Downdraft' gasifiers can minimise such problems and also cut down on the pollutants such as PAH, in the exhaust from the electricity generator.

Many developing countries have small fossil fuel reserves and so they could profitably save the scarce hard currency needed to buy energy imports by growing their own crops for energy purposes.

Making Ethanol from Biomass

As all home-brewers know very well, biomass material can be fermented with yeast to produce ethyl alcohol, or ethanol. Brazil uses part of its sugar crops to manufacture around a fifth of Brazil's fuel transportation needs. The bagasse wastes that arise from this process are used to produce large amounts of heat and electricity for industry. In all, a total of 12 per cent of Brazil's energy comes from sugar in one form or another.[35]

The fuel-alcohol programme was started in earnest in 1975 with a series of incentives. The Brazilians were desperate to cut down on an oil import bill that had jumped massively upwards and which had caused an acute shortage of hard currency. Since the 1970s the programme has been an economic success. By 1993 the cost of producing ethanol had fallen by nearly a half compared with the late 1970s (from $70 to $40 per barrel equivalent) and a further reduction in price to $32 per barrel equivalent is expected within a few years.[36] The industry now no longer receives significant incentives.

This cost-cutting process is likely to continue with the development of more productive and pest-resistant cane, the increasing use of mechanisation in harvesting, and better recycling of wastes.

The biggest problem faced by the industry has been its inability to expand quickly enough to satisfy demand. The sales of cars that run on neat alcohol increased dramatically at the end of the 1980s. The Brazilians are hoping to secure an assured market in the US where there is a small, but growing, move towards running cars on a blend of ethanol and gasoline.[37]

Ethanol produced from feedstocks such as corn grown in temperate regions is much less economic than production from sugarcane in the tropics. The amount of biomass produced from a given area per year is much lower in the case of corn grown in cool climates. Indeed, the energy used to produce the ethanol from corn in such conditions can rival the energy content of the ethanol fuel product. The degree to which carbon dioxide emissions can be cut by substituting ethanol grown from corn in temperate zones for gasoline is probably modest. However, the production of ethanol from otherwise waste agricultural production is a different matter.

Even in Europe and the US it is still economic to use biomass wastes for ethanol production. Indeed, there are suggestions that this end could best be achieved by using a bacillus to break down the hemicellulose in the wastes.[38]

Anaerobic Decomposition

Anaerobic decomposition involves the action of bacteria to decompose organic materials in the absence of oxygen. Methane and carbon dioxide are produced. This is called biogas and it can be used as an energy source.

Anaerobic decomposition occurs 'unintentionally' in landfill sites, and the methane gas can present a major hazard. In recent years it has been the practice to capture some of these emissions (often referred to as 'landfill gas') and use them to generate electricity or provide heat for some local purpose. The process of anaerobic decomposition can also be used 'intentionally' in specially constructed anaerobic digestion tanks for several purposes.

The first is in the treatment of animal and human wastes (sewage). The second is in the treatment of wastes from the food-processing industries (wastewater treatment). With the tightening of environmental regulations concerning the disposal of wastes of various sorts, there is a market niche for anaerobic digestion techniques. These techniques can be used to turn the wastes into fertiliser and, simultaneously, to produce biogas.

In the UK around 70 MW of landfill gas and about 30 MW of sewage gas projects had been commissioned by September 1993 under the UK's renewable energy programme.[39]

The cost of biogas is usually too high for it to be considered as an economic alternative to natural gas in those circumstances where natural gas is readily available, but the biogas can be a valuable by-product of the process of treating organic waste materials.

Biogas can be economically viable, as the main product, in those situations where conventional energy supplies are not readily available. Agricultural communities in developing countries which are not connected to electricity grids or natural gas pipelines but which have plenty of animal dung often have the potential to make effective use of biogas technology. Biogas systems can be deployed either on a family or a communal basis, although communal schemes can be the most cost-effective. The projects can be built using local labour resources, although the development of such schemes depends on the existence of a large pool of expertise and training.

Water Power

Hydroelectricity

Hydropower, the conversion of the energy in moving water to mechanical or electrical energy, has been used for more than 2,000 years. Hydroelectricity involves use of a turbine designed to be driven by water. Early electricity schemes were frequently hydro-based. Hydroelectric schemes provided around 18 per cent of world electricity in 1990.

The size of the hydropower resource is determined by the amount of precipitation. In theory the total hydro resource could supply world electricity demand several times over.

Hydro schemes come in many shapes and sizes. The largest quantities of power are generated from schemes with high heads (water falls through a large height) and with large volume flows of water. High heads will most often occur in mountainous regions where dams are built across narrow river valleys. Reservoirs, which help to even out the variability of river flows and can provide flood control or irrigation services, are formed behind the dams. Nevertheless, production varies with the seasons so capacity factors are smaller than average fossil-fuelled plant.

Reservoir systems can be used to provide so-called pumped storage whereby energy can be stored through pumping water uphill at periods of slack demand, to be released to produce electricity at short notice at times of peak demand.

The Itaipu scheme in Brazil is the largest hydro-scheme in the world. At 12,600 MW its total capacity is equivalent to about a dozen large nuclear power stations. At the other end of the scale there are schemes of just a few kilowatts; these are a few of the options for supplying electricity to communities too remote to be connected economically to the grid.

In industrialised countries, in particular Sweden, Norway and Switzerland, a large proportion of hydroelectric potential has already been tapped. Around 10 per cent of US electricity supply comes from hydropower, much of it built by public utilities like the Tennessee Valley Authority. Little more than 1 per cent of the UK's electricity comes from hydropower, although much potential exists in Scotland.

The prospects for large increases in hydropower capacity in the industrialised world are dim chiefly because of environmental opposition, although there is some growth in small schemes and some upgrading of existing facilities. Several rivers in countries like the US and Sweden are protected by law from hydro development.

There is likely to be continued major development of hydro-electric power in the developing world. In China a massive 42,000 MW of hydro capacity is projected in the 1991 to 2000 period.[40] Hydropower resources are still relatively underused in many developing countries and the environmental pressures against hydropower developments are much weaker than in the West, while the pressure for more energy production is much higher.

The biggest environmental impact of large hydroelectric schemes results from the inundation of large areas of land by the reservoirs that form behind the dams. These can displace many people, destroy fauna and flora, and may create stagnant aquatic conditions conducive to the spread of malaria and bilharzia.

One of the most controversial schemes is the proposed Three Gorges dam across the Yangste in China. The projected capacity of the project is at least 13,000 MW. This would make it the largest hydro project ever built. It involves the creation of a 500-km-long lake and the displacement of over a million people as 140 towns disappear.[41]

The World Bank is not involved in this scheme but it has been forced to pull out of other projects such as the Sadar Sarovar project in the Narmada Basin of the Indian state of Gujarat. This 1,450 MW project would displace nearly 70,000 people.

The forced resettlement of thousands of people has led many environmentalists to denounce the World Bank and other aid institutions for supporting hydroelectric schemes. In 1994 a fierce row blew up over British aid given to the Pergau Dam project in Malaysia. It has been alleged that funding of the dam had been linked to British arms sales to Malaysia.

There have been many complaints about the impact of existing hydroelectric schemes on fish, especially species of fish that migrate. For example, around 16 million salmon used to migrate along the Columbia River in Washington State to the ocean. Attempts have been made to breed smolts (young salmon) in specially organised hatcheries. The smolts are then transported around the dams. However, now only 2 million fish migrate.[42]

The costs of hydropower schemes vary enormously. The majority of the costs are usually concerned with the civil works. Running costs are very low. Hydropower plant will have a much longer lifetime than most other energy projects. Assuming $1,500 per kW, a 40 per cent capacity factor and a 10 per cent discount rate, then the cost of electricity will be around 4.8 c/kWh for a 50-year lifetime, rising to about 6 c/kWh for a conventional 15-year contract.

Usually only public sector funding can deal with the costs of financing over the long term, which is why hydropower projects are usually financed by government agencies.

Small schemes can have relatively low costs. The projects given contracts under the UK Renewable Energy Programme (average size under 1 MW) in 1991 were financed (privately) at 6 p/kWh on a six-year contract which equates to just under 3 p/kWh over 50 years. Smaller projects will have less environmental impact than larger ones, although there will still be controversies about fish and other issues.

Tidal Power

Tides are caused by the gravitational effects of the Sun and the Moon on the oceans. The rising and falling effect of the tides can be best exploited when river estuaries or bays accentuate the height of the tide.

Water turbines are used to turn the potential energy of the water into electricity. The turbines work as the tides rise, and then, after the sluice gates have shut trapping the high tide behind the barrage, the water passes through the turbines in a seaward direction.

The biggest tidal device built so far has been the 240 MW scheme at La Rance in France. Smaller devices have been constructed in China, Canada and Russia.

It has been estimated that around 8 per cent of all world energy supplies would be met if all the tidal power available for less than 20 c/kWh was exploited. The largest tidal schemes are likely to be at Penzhina, Mezen and Tugur in Russia, the San Jose Gulf in Argentina, the Severn Estuary in the UK, and Turnagain Arm in the US.[43]

Tidal power is expensive if viewed solely as a power source. The civil engineering works are costly, there is a relatively low 'head' and the tides rise and fall only twice a day. This latter factor leads to a very low capacity factor. On the other hand, tidal barrages will last a long time and can have 'developmental' advantages, including the construction of roads, the creation of marinas and ports, and other things. They are likely to be funded by public rather than private bodies.

The costs of the River Severn barrage (which would provide 6 per cent of UK electricity demand) would be almost 10 p/kWh at a 10 per cent discount rate (assuming a 120-year lifetime) and about 6 p/kWh if a lower, 5 per cent, discount rate is used to take account of other benefits.

Even though tidal schemes cause much less environmental impact than hydro projects, they still have important critics. In the UK the Royal Society for the Protection of Birds is strongly opposed to them on account of the possible loss of mudflat habitats for wading birds. The RSPB is the largest pressure group in Britain.

Tidal Streams

Tidal stream power technology basically involves use of 'underwater windmills' in strong tidal stream currents, for example in between islands and the coast. The tidal turbines (or tidemills) could be tethered by ropes tied to anchors on the sea bed. Such systems would not require the extensive civil engineering works that go with tidal barrages, although the energy available to the tidemills would be much less than that available to the barrage turbines.

Some research into tidemills is being conducted in the UK and Spain. However, these are only small programmes and without much more substantial governmental help it will be many years before it is possible to evaluate properly what the contribution of tidemills could be, never mind make some firm statements about the economics of the technology.

Wave Power

Interest in wave power goes back no further than the onset of the 1973 oil crisis, and most research into it has been conducted, perhaps unsurprisingly, by countries with long coastlines. The UK, Norway and Japan have been the leaders in this technology, although many others ranging from China to the US are also doing some work.

The North Atlantic, which is at the end of a 'wave stream' starting in the Gulf of Mexico, has the highest measured average wave energy. In theory the UK could derive all of its energy from wave power, although the practical resource has been described as being nearer about a quarter of the UK's electricity demand.[44] The West Coast of the US may also be particularly suitable for wave power.

Wave power machines can be mounted at the shore or offshore. So far the cheapest devices, like the 350 kW Norwegian 'tapered channel' scheme sited near Bergen, have been onshore devices. However, the onshore resource is relatively small because waves lose much of their energy as they meet the shoreline and many sites are ruled out because the large size of the tidal range poses problems for the collection devices. Nevertheless, onshore devices in relatively stormy parts could provide energy to remote regions at about 6 p/kWh (8 per cent discount rate, 1992 prices).[45]

There is a multiplicity of designs for offshore devices. These can capture more wave energy and the resource is potentially large, but the maintenance costs will be higher than in the case of onshore machines.

The so-called 'oscillating water column' design, which drives a turbine by air compression, is currently a favourite for demonstration devices. This includes the largest wave power device planned so far, the 2 MW so-called 'Osprey' (Ocean Swell Powered Renewable Energy) machine. This would be sited off the Scottish coast near Dounreay.[46]

The UK, which is better placed than most countries to exploit wave power, has nevertheless withdrawn funding from wave power research programmes on the grounds that projected costs are too high and export opportunities too small. Wave power is likely to perform best in cold, choppy, waters, but energy can also be drawn from the ocean in much warmer climes. Norway is exporting wave power machines to the Pacific.

OTEC

OTEC stands for Ocean Thermal Energy Conversion. It involves using the temperature gradient between surface and colder, deeper, waters to drive a turbine. Various experimental schemes have been mooted in a range of tropical locations (where the temperature gradients are the greatest) including a US-backed 100 MW proposal in Puerto Rico. However, since the early 1980s it has been difficult to raise the funds needed to turn feasibility studies into reality.

The main problems faced by engineers are the logistics of dealing with the massive flows of water needed to provide the energy. As yet the problems associated with running such a plant on a long-term basis have not been overcome. However, if they were, then the resource would be massive, limited only by the difficulties of transmitting the electricity across long distances to the markets. In practice OTEC will be most appropriate for tropical islands.

Geothermal Energy

Geothermal energy is associated with the internal heat of the Earth which is partially maintained by natural radioactivity. Geothermal energy could therefore be regarded as passive nuclear power.

There are many millions of times as much heat inside the planet as has ever been used by humans throughout its history. For practical purposes the resource is limited to the first few kilometres underground, but this still leaves a resource many thousands of times bigger than current energy requirements.

The most accessible form of geothermal energy is called hydrothermal because its comes from natural aquifers. These are hottest in tectonically active areas of the world such as Western US and Italy (and indeed in a significant portion of the world as a whole). Resources exist elsewhere, for example a small hydrothermal scheme exists in Southampton in the UK to supply local heating services, but the higher temperatures come from areas with volcanic activity.

Geothermal resources are not usually well situated for serving heating needs, but if temperatures are high enough, electricity can be generated using steam turbines and the power can be transported through the wires. Although there has been geothermal electricity production since the Italians started in 1904, it only really took off after the oil crises. Recently geothermal electricity expansion has slowed, but there are still many plans for schemes around the world.

A sizeable portion of the world's geothermal electricity capacity is at The Geysers in Northern California and around 6,000 MW of capacity exist worldwide, as can be seen in Table 12.1.

Table 12.1 Global Geothermal Electricity Production (MW capacity for 1992)

US	2,700
Philippines	891
Mexico	700
Italy	545
New Zealand	283
Japan	270
Indonesia	142
El Salvador	95
Keyna	45
Iceland	45
Nicaragua	35
Turkey	21
China	21
CIS	11
France	4.2
Portugal	3
Guatemala	2
Thailand	0.3
Total	5,813.5

Such schemes can be cheap. A new project at Meager Creek in British Columbia will come in at around 4.5 c/kWh (10 per cent discount rate, 15-year contract) and for less than 4 c/kWh when upgraded to 500 MW.[47] Many projects cost more than this, but these are usually in places where the cost of supplying power from other sources would be very high.

In fact, hydrothermal energy resources could, like natural gas resources, be very much larger than at first thought. People have not been looking very much. Given institutional encouragement, hydrothermal energy could be a significant contributor to world electricity supplies in the medium term.

Geothermal energy's environmental impacts are concerned with emissions like hydrogen sulphide gas. These can be re-injected into the ground. There have been complaints about the drilling and setting up of some schemes, especially in the Philippines.

Hot dry rock (HDR) geothermal energy is as yet an under-developed technology. HDR can be tapped by creating and then drawing on reservoirs of water in existing hot rocks. Resources are enormous, but despite research at places like Fenton Hill in the US, Cornwall in the UK and also in Japan, many engineering problems remain to be solved.

Another potentially large resource is geopressured brine. Hot liquids carrying large amounts of methane could be exploited both for the hydrothermal and the methane energy content. There are known to be considerable resources around the US coast of the Gulf of Mexico. However, the brine has a very corrosive effect on metals and advances in materials technology are necessary before this technology can be successfully demonstrated.

Geothermal energy generally, unlike many other renewable energy sources, is not intermittent and its plant has a high capacity factor, although the steam turbine technology it uses makes for relatively inefficient conversion of steam into electricity. Other renewable energy sources may have more efficient conversion systems, but they are often intermittent. They need storage facilities to allow them to supply continuous energy needs. Hydrogen is a possible way of storing renewable energy.

Solar Hydrogen

The idea of solar hydrogen has been introduced by authors such as Ogden and Williams.[48] Renewable electricity sources like wind power and solar PV could, in theory, be used to electrolyse water to produce hydrogen (and oxygen). The hydrogen could then be transported along pipelines to be used in place of natural gas. It could also be a fuel for fuel cells in uses such as motor transport.

Hydrogen is a very clean fuel, with low NOx, carbon monoxide emissions and so on, even compared with natural gas. The problems of transporting and storing the gas are, in principle, not wholly different from methane, which is itself a highly inflammable substance. This suggests that some time in the future hydrogen may act as an important energy carrier.

There are at present only a few specialised markets for hydrogen in industry, but there are possibilities for using hydrogen as a fuel supply in places like California where there are especially acute air-quality problems. Of course this does not necessarily mean that the hydrogen will be produced from renewable energy; at the moment fossil fuels provide the cheapest feedstocks for hydrogen production. The costs of renewable fuels will have to decline considerably before they can be cost-effectively used to produce hydrogen.

Solar hydrogen is a concept whose day will come. When it does, it may just as easily be in the context of deriving hydrogen from very cheaply produced biomass material that has been gasified as from solar PV or wind power.

Another more long-term possibility is a process called photolysis. This is the use of sunlight to split water directly into hydrogen and oxygen. Photolysis of water is done only experimentally and the efficiency of the process will have to be improved greatly, perhaps through the use of some form of catalyst, before it is of even remote commercial interest.

Interest in hydrogen as a fuel has partly been driven by the perceived need to find a storage medium for renewable energy systems. However, this discussion may be largely irrelevant to medium-term realities. For a start, energy storage systems (whether flywheels, pumped storage facilities, substances that change phases, or use of hydrogen) add to the cost of energy systems. Second, various studies have suggested that total electricity supplies from grid systems could include contributions of more than 20 per cent from single intermittent sources without financial penalty, and we are a long way from seeing this proportion being reached. Third, some renewable energy sources, notably biomass, can be stored with relative ease.

The Future of Renewable Energy

Renewable energy systems have often been touted as a means of eliminating the need for electricity grids. While there are a growing number of important instances where renewable energy sources can supply energy in off-grid situations, there are also many bountiful opportunities for supplying renewable energy to the grid. Indeed, grid supply enables renewable energy technologies to enjoy many cost advantages.

Connection to the grid means that renewable energy systems do not have to be weighed down by the cost of expensive energy storage systems and any intermittence can be absorbed by the grid.

A second advantage of grid connection is that the best sites for renewable energy production (which are often remote from large groups of consumers) can be selected and the power can then be fed into the grid.

A third advantage is that sets of equipment are usually a lot cheaper if ordered in large batches, and wind farms consisting of many wind turbines or large new housing developments including PV panels can realise the discounts that go with batch production. Grid supply is likely to present many more opportunities for such (relatively) larger schemes than isolated homes or communities. Of course, as was discussed earlier, there are many situations where the lower cost of large wind farms is offset by their planning disadvantages. This situation may favour smaller though still usually grid-connected schemes financed by local communties.

Indeed, if renewable energy is to be a central rather than a relatively marginal player in the energy world, it has to grow big in the grid-supplied market.

Although there are markets where renewable energy systems are already competitive with other energy sources, many renewable energy technologies need to be given 'protected' sectors in energy markets if they are to develop. In time various renewable energy technologies will be able to compete in terms of costs with fossil fuels, but until then financial support will be needed.

There is no substitute for commercial deployment of new renewable energy technologies. Engineering advances are made by the attrition of commercial application, not just by boffins having bright ideas in laboratories.

There is widespread support for renewable energy technologies to be given a good chance. From Amsterdam to Los Angeles there are suggestions that individual consumers might be given the option of paying a bit extra for their electricity in order to support emerging renewable energy technologies.

The payment of high prices for renewable energy subsidised by electricity consumers rather than taxpayers is becoming the preferred form of support. Competitive bidding systems are being used to award contracts in California and the UK.

In California half of all new electricity supplies are now coming from renewable energy, mainly wind power and geothermal energy, and the other half from natural gas cogeneration.

Renewable energy has long-term economic advantages over fossil fuels. Energy from fossil fuels is cheap only because, despite the costs of supplying fossil fuels, low-cost technologies have matured which convert and deliver the energy with high degrees of efficiency. Renewable energy technologies, whose development is motivated by a desire to reduce energy dependence and reduce pollution from fossil fuels, will also mature.

If renewable energy technologies are given institutional and financial support in the short term, then the difference between the costs of fossil fuel and renewable energy technologies will become increasingly smaller.

Several renewable energy sources are improving the efficiency with which they convert solar energy into delivered energy and they are lowering their capital costs at a rapid rate. The capital costs of mature fossil fuel technologies are, by comparison, likely to decline at a much more gentle rate. Decentralised renewable energy sources involve small units, the capital costs of which will be greatly reduced if they are mass produced. Large power stations do not enjoy such benefits and they take much longer and thus cost more to install than decentralised renewable energy systems.

Fossil fuel prices are, in the context of increasing energy insecurity and longer supply lines, unlikely to decline in the long term. Thus the low or zero cost of fuel for renewable energy sources is likely to give renewable energy technologies an ever more important edge over their fossil fuel competitors as the capital costs of renewable energy systems decline, their efficiencies improve and their markets grow. In short, renewable energy will ultimately displace fossil fuels as the major energy source because renewable energy will become cheaper than energy derived from fossil fuels.

The costs of several forms of renewable energy – some types of biomass, hydro schemes and geothermal power – are competitive already. Wind power is rapidly heading in that direction and solar PV systems are likely to become much cheaper as their penetration of a widening number of markets increases.

In the case of nuclear power there are few possibilities for increasing conversion efficiencies since nuclear power is reliant on steam-raising technology. The problems of nuclear safety are likely to preclude any significant reductions in capital costs. Arguments about the environmental acceptability of nuclear waste are likely to continue.

There are controversies about the environmental impacts of renewable energy. Some types, like large hydro schemes, are extremely intrusive but many others are much less polluting than most commercial and agricultural activities. True, some renewable energy technologies are land-intensive, but by contrast the impacts of fossil fuels are far reaching and global even though many of their impacts are very remote from the energy consumer. This is one of the problems with fossil fuels. Their supply lines are often long and they are becoming longer. In the instance of renewable energy it is a case of 'what you see is what you get'.

Planning problems may slow the deployment of some types of renewable energy in the most densely populated regions of Western Europe, but even here opportunities for local ownership of the projects can win many hearts and minds. As renewable energy becomes cheaper the commercial rationale for these fuels will increase and this may offset many of the planning problems.

In reality there are many cards in the renewable hand. Some of them are rapidly improving in value and could provide some big surprises in the relatively near term.

13

A Low Cost Planet?

An important aim of this book is to examine the established view that solving fundamental energy–environmental problems inevitably increases the monetary costs of supplying energy services.

'End of pipe' anti-pollution measures do increase costs, although by smaller amounts than often claimed by energy supply interest groups. Even here stricter emissions standards often give a boost to relatively clean fuel technologies, such as natural gas, which are already cheap.

I have shown that quite radical improvements in energy efficiency can be and are being achieved for no extra short-term (or long-term) cost to the consumer. Such improvements are an effective way of abating conventional problems of energy policy such as energy dependence. Energy prices are held down in the process.

Clear evidence is emerging that energy efficiency programmes, such as those pursued in The Netherlands and, to a certain extent, in some parts of the US, are achieving impressive reductions in carbon dioxide emissions and reductions in emissions of other pollutants in the context of lower energy bills for the consumer.

Improvements in the effectiveness with which energy efficiency programmes are delivered, declining costs of energy efficiency techniques, increased investment in energy efficiency and a general trend towards the production of goods and services with relatively low energy inputs could accelerate these achievements.

The biggest obstacles to energy efficiency are institutional, not economic or technical. Institutions can be changed. There is every reason to believe that they will be changed in the future.

The spread of techniques such as demand-side management and high overall energy efficiency cogeneration has so far been slow. Much of the sloth can be explained by the power of energy supply interest groups whose resources and entrenched political clout give them tremendous opportunities to generate misinformation and delay the adoption of energy-efficient practices.

Progress has also been slowed by the reluctance of opinion formers, in particular many economists, to understand that the advantages of energy efficiency can be optimised only by intervention in energy markets to create markets in energy efficiency. Free market economists will find that their credibility will suffer if they damn systems such as integrated resource planning as 'bureaucratic' without proposing low-cost measures that will ensure that the amount of investment in energy efficiency is massively increased. There are several means of assuring this expansion of energy efficiency investment. Effective programmes to increase investment in energy efficiency will act to increase competition in markets for energy services.

The traditional energy suppliers and their advocates have insisted that environmentalist objectives, such as absolute reductions in carbon dioxide emissions, can be achieved only by increasing the costs of supplying energy to the consumer. This is a straw man. Yet many environmentalists who have called for high energy taxes have, in effect, been auditioning for the part!

It is unlikely that voters will sanction large increases in energy taxes purely for the reason of cutting carbon dioxide emissions while there is the possibility that emissions can be cut using means that reduce, not increase, energy bills.

The demand for high energy taxes distracts attention away from far cheaper, often regulatory, means of accomplishing improvements in energy efficiency. Increased investment in energy efficiency is the crucial objective, not increased energy prices.

Environmentalists will win the argument with the conventional energy suppliers. However, they will do so by making common cause with consumers in reducing their bills rather than incurring their hostility through making demands for massive tax increases. In fact, environmentalists have a much clearer claim to be on the side of consumer interests than energy supply interest groups. The energy suppliers want to supply as much energy as possible at the highest price. Environmentalists want to reduce energy consumption, an aim that can save money, help to keep energy prices low and reduce pollution into the bargain.

Sustainable energy strategies are cheap, clean and modern. By contrast conventional approaches are expensive, dirty and outdated.

The convergence of environmentalist and consumer interests is underscored by the shape of technological development. This favours low-energy strategies. The spread of information technology increases the ability to design, monitor and control energy flows. This increases the potential for, and decreases the costs of, energy efficiency techniques. Much future economic growth will consist of 'value added' and information technology services rather than traditional heavy industry, a trend which reduces the use of energy in industry.

A further drive towards environmentalist objectives is created by cultural changes. Society is now much more receptive to the idea that high living standards can be and ought to be raised through improving resource efficiency and by using renewable rather than finite resources.

Nuclear power used to be be almost universally seen as the future solution to problems caused by fossil fuels. These days there is much less enthusiasm. Nuclear power is very expensive compared with many energy efficiency techniques and also, increasingly, renewable fuels, despite the latter's relative state of underdevelopment. Nuclear power is also in a technological cul-de-sac. Nuclear technology, with its complicated valve systems, its expensive safety features and steam generators seems to be more typical of mid-twentieth century heavy industry rather than light, decentralised, twenty-first century information-centred industry. The apparently insoluble problem of radioactive nuclear waste reinforces nuclear power's image as a dirty sunset technology.

Renewable energy technologies are usually more at ease with 'environmentalist' culture than nuclear power, although sometimes environmentalists themselves spearhead opposition to renewable energy developments. However, given that all energy sources have environmental impacts, greens will look unrealistic if they oppose practical near-term efforts to develop and deploy renewable energy technologies. Energy efficiency is of fundamental importance, but everybody knows that we must derive energy from somewhere.

Clearly, if both global fossil fuel use and the world's rate of carbon dioxide emissions are to be greatly reduced, and not just contained, then we will need to use larger amounts of renewable energy as the twenty-first century progresses. Unless such technologies are given an early boost then this may not happen, with costly consequences.

The cost of renewable energy technologies is falling rapidly. The often decentralised nature and low or zero fuel costs of renewable energy technologies give them inherent advantages over conventional fuels. In the medium and long term several will be at least as cheap, or even cheaper, than energy derived from fossil fuels.

An effective renewable energy programme is a much cheaper and a much surer path to lowering energy costs in the future than giving subsidies to nuclear power.

The traditional energy suppliers try to characterise the green lobby as being against economic development. It is true that some ecologists are disdainful of talk of 'development'. Yet, if the evidence and arguments in this book are at all accurate it is likely that sustainable energy strategies will succeed in enhancing, not reducing, the rate of economic development in 'industrialised' and 'developing' countries alike. This fits in with the wishes of most ordinary people (in both rich and poor countries) for higher material living standards.

Ultimately the green message is likely to spread in the developed and the developing world because ecologists will explain how sustainable energy policies and other environmentally sensitive strategies will assist economic development and the achievement of environmental objectives at the same time.

This conclusion rests upon the assumption that the green movement can mobilise consumers and voters more effectively in support of sustainable energy strategies than it has done in the past.

So far there have been two waves of ecological consciousness in the West which have both served to make green issues a more important part of culture. The first wave peaked in the early 1970s. A central theme of this first wave was the idea that pollution problems may limit the possibilities for economic growth.[1] The second wave peaked at the end of the 1980s. The second wave's theme was less pessimistic and documents such as the Brundtland Report[2] talked of the concept of sustainable development, development that accounted for and minimised environmental costs. Each wave took ecological issues further up the political and cultural agenda.

In a few years' time there may be a third wave of ecological consciousness. This time a central theme might encapsulate the idea that environmentally sensitive strategies can act to improve economic development even before account is made of the external costs of economic development.

The idea that fundamental energy–environmental problems such as resource depletion and global warming can be solved through measures that involve no extra costs to the consumer is likely to grow much stronger.

Notes

2. The Pollution Problem

1. NOx is emitted in the form of NO, NO_2 and N_2O, respectively called nitric oxide, nitrous oxide and nitrogen dioxide. The NO is rapidly oxidised to produce NO_2 which forms the bulk of NOx deposition.
2. J. McCormick, *Acid Earth* (London: Earthscan, 1989).
3. 1990 Forest Damage Survey in Air Pollution Studies No. 8, *Impacts of Long Range Transboundary Air Pollution* (New York: United Nations, 1992).
4. K.A. Gourlay, *World of Waste* (London: Zed Books, 1991), p. 162.
5. J. McCormick, *Acid Earth*, pp. 186–7.
6. Environmental Protection Agency, *Progress Report on US/Canada Air Quality Agreement* (Washington DC, 1992).
7. W. Bown, 'Dying from Too Much Dust', *New Scientist*, 12 March 1994, pp. 12–13.
8. World Commission on Environment and Development, *Our Common Future* (Oxford: OUP, 1987)
9. UNEP/WMO IPCC, B. Callender, J. Houghton and S. Varney (eds), *Climate Change 1992, Supplementary Report to the IPCC Scientific Assessment* (Cambridge: CUP, 1992), p. 91.
10. T.M.L. Wigley and S.C.B. Raper, 'Implication for Climate and Sea Level of Revised IPCC Emissions Scenarios', *Nature*, Vol. 357, No. 6376, May 1992.
11. S. Ryan, 'Scientists Dismiss Global Warming Leading to Floods', *Sunday Times*, 27 March 1994, p. 5.
12. N.J. Shackleton and N.G. Pisias, 'The Carbon Cycle and Atmospheric Carbon Dioxide: Natural Variations Archean to Present', *Geophysical Monograph* 32 (American Geophysical Union, 1985).
13. E. Friis-Christensen and K. Lassen, 'Length of the Solar Cycle: an Indicator of Solar Activity Closely Associated with Climate', *Science*, Vol. 254, pp. 698–700.

14. P.M. Kelly and T.M.L. Wigley, 'Solar Cycle Length, Greenhouse Forcing and Global Climate', letter to *Nature*, Vol. 360, 26 November 1992.
15. F. Pearce, 'American Sceptic Plays Down Global Warming Fears', *New Scientist*, 19 December 1992, p. 6.
16. By a sad coincidence Exxon had portrayed in their corporate calendar 'a sunlit *Exxon Valdez* sailing through a pristine Cook Inlet in the very month of the year that the unfortunate vessel collided with the said Cook Inlet' ('FGD and All That', *Energy Economist*, January 1994, p. 7).
17. M. Hornsby, 'North Sea Birds Thrive Despite Oil Spills', *The Times*, 25 March 1993, p. 4.
18. National Research Council, 'Using Oil Dispersants on the Sea' (Washington: National Academy Press, 1989, pp. 9–11).
19. 'The Wreck of the Braer', *The Economist*, 9 January 1993, pp. 25–6.
20. National Research Council, 'Using Oil Dispersants', p. 318.
21. N. Hawkes, 'Wildlife Experts in Dispute over Safety of Oil Dispersants,' *The Times*, 8 January 1993, p. 3.
22. D. Pallister, 'Nigerian Tribe puts Environment on Election Agenda', *Guardian*, 17 May 1993, p. 9.
23. F. Pearce, 'The Scandal of Siberia', *New Scientist*, 27 November 1993, p. 33.
24. See D. Maclaine, 'Fields of Doubt', *Electrical Review*, 13 May 1994, pp. 28–30.
25. N. Nuttall, 'Scientists Doubt Cancer Link with Electricity', *The Times*, 1 April 1992, p. 6.
26. S. Spiller and T. Turner, 'Hidden Powers', *Guardian* supplement, 8 April 1994, pp. 16–17.
27. 'Ray Ipsa Loquitor', *The Economist*, 6 March 1993, p. 87.

3. The Resource Problem

1. C. Ponting, *A Green History of the World* (Harmondsworth: Penguin Books, 1992).
2. See 'The 1996 Oil Shock', *Energy Economist*, May 1993, pp. 17–23 for a well-written summary of these arguments; and D. Thomas Gochenor, 'The Coming Capacity Shortfall', *Energy Policy*, October 1992, pp. 973–82 for an analysis of problems facing OPEC attempts to expand supply capacity.
3. P. Odell, 'Prospects for Non-OPEC Oil Supply', *Energy Policy*, October 1992, pp. 931–41.

4. Personal communication from Paul Horsman, Greenpeace International, London, September 1993.

4. Solutions

1. A. Lovins, *Soft Energy Paths* (Harmondsworth: Penguin Books, 1977).

5. Costing Solutions

1. Many of these methods are summarised in D. Pearce, A. Markandya and E. Barbier, *Blueprint for a Green Economy* (London: Earthscan, 1989); and, in the same series, D. Pearce (ed.), *Blueprint 2, Greening the World Economy* (London: Earthscan, 1991).
2. See V. Anderson, *Energy Efficiency Policies* (London: Routledge, 1993); and M. Jacobs, *The Green Economy* (London: Pluto Press, 1991), for reviews of monetary valuation techniques, cost-benefit analysis and related issues.
3. A. Stirling, 'Regulating the Electricity Supply Industry by Valuing Environmental Effects', *Futures*, December 1992, pp. 1024–47.
4. See, for example, W. Nordhaus, 'How Fast Should We Graze the Global Commons?', *American Economic Review, Papers and Proceedings*, Vol. 72, No. 2, pp. 242–6.
5. One example among many is the extra cost of meeting the tight low-sulphur diesel fuel regulations in California. Diesel prices have increased by 6 cents per gallon, half the additional costs calculated when the rules were originally adopted. See *Oil & Gas Journal*, 22 May 1993, pp. 34–5.
6. Knowledge could be considered as a factor of production alongside the traditional ones of land, labour, capital and enterprise. It is a resource that is increasing in importance. See interview with Alvin Toffler in *New Scientist*, 19 March 1994, pp. 22–5.
7. The principal exception to this rule is the case of low-income consumers living in very old houses where insulation measures may enable them to keep warm for longer periods while consuming the same amount of energy. In other circumstances, for example refrigeration, lowering the cost of providing an energy service by improving energy efficiency will not generally encourage the consumer to use more energy.
8. Stirling, 'Regulating the Electricity Supply Industry'.

6. Gas, Gas and More Gas?

1. The classification of the reasons for the advance of natural gas into 'technological', 'economic' and 'environmental' categories is taken from J. Homer, *Natural Gas in Developing Countries* (Washington DC: World Bank, 1993).
2. Oil produces around 80 per cent of the carbon dioxide emitted when coal is burned to produce an equivalent quantity of energy. These figures take no account of the technology used to burn the fuel.
3. Homer, *Natural Gas*, p. 24.
4. Ibid., p. 26.
5. J. Kolodziejski, 'New Supplies Help Boost Industry Confidence', *Gas World International*, August 1993, pp. 11–12.
6. A major factor in this change appears to have been the desire of newly privatised regional electricity distribution companies to invest in 'independent' CCGT projects to give them alternative sources of power to those (mostly coal-fired sources) supplied by the central electricity generators, National Power and PowerGen.
7. 'Industry Must Avoid Gas vs Oil Civil War', *Oil & Gas Journal*, 4 January 1993, p. 17.
8. P.J. Crutzen, 'Methane's Sinks and Sources', *Nature*, Vol. 350, pp. 380–1.
9. *Energy Economist*, April 1993, p. 32.
10. See, for example, M. Prior, *European Energy Policy and Environmental Objectives in the 90s* (York: Planning for Energy and the Environment, 1992).

7. Energy Efficiency

1. Danish Energy Agency, *Energy Efficiency in Denmark* (Copenhagen, 1992), p. 22. Also personal communication from P. Knudsen of Nellemann cogeneration consultants, Aalborg, Denmark.
2. Hans von Bulow (Danish Energy Ministry) in evidence to House of Lords Select Committee on European Communities, *13th Report 1990–1991* (London: HMSO, 1991), p. 129.
3. Personal communication from Tom Thompson, California Public Utilities Commission.

4. See R.D. Evans, *Environmental and Economic Implications of Small-scale CHP* (Harwell: Energy Technology Support Unit, 1990).

5. 'US to Get Molten Carbonate Fuel Cell Plant', *Electrical Review*, 15 October 1993, p. 19.

6. Cited in W.C. Patterson, *The Energy Alternative* (London: Boxtree, 1990), p. 97.

7. World Bank, *Energy Efficiency and Conservation in the Developing World* (Washington DC: World Bank/IBRD, 1993), p. 31.

8. See *Retrofitting AC Variable Speed Drives* (Energy Efficiency Office/ETSU, 1991).

9. D. Olivier, *Energy Efficiency and Renewables: Recent Experience on Mainland Europe* (Credenhill, Herefordshire: Energy Advisory Associates, 1992).

10. World Bank, *Energy Efficiency*, 1993, p. 51.

11. Electricity producers sometimes claim that electric heating will be provided by 'low emission' non-fossil and natural gas plant in off-peak periods. In fact the extra power is likely to come from coal-fired electricity power stations and the resultant heating is associated with a great deal more pollution than natural gas heating.

12. R. Webb, 'Offices that Breathe Naturally', *New Scientist*, 11 June 1994, pp. 38–41.

13. See, for example, T. Woolf and E. Derosa Lutz, 'Energy Efficiency in Britain', *Energy Policy*, July 1993, p. 236 and p. 241, note 17.

14. E. Mills and J. Aizenberg, 'Lighting for Glasnost', *International Association for Energy Efficient Lighting Newsletter*, January 1994, No. 5, Vol. 2, pp. 8–10.

15. *The Economist*, 5 December 1992, pp. 122–3; and *ENDS Report 221*, June 1993, p. 25.

16. D. Olivier, 'Energy Efficient Office', *Safe Energy*, October/November 1993, pp. 12–13.

17. L. Schipper and D.V. Hawke, 'More Efficient Household Electricity-use', *Energy Policy*, April 1991, pp. 244–65.

18. D. Olivier et al., *Energy Efficient Futures* (London: Earth Resources Research, 1983).

19. H. Geller, *National Energy Efficiency Platform* (Berkeley, CA: American Council for an Energy Efficient Future, 1989).

20. *Protecting the Earth*, 3rd Report of Enquete Commission (Bonn: Economica Verlag, 1991).

21. See V. Anderson, *Energy Efficiency Policies* (London: Routledge, 1993), for a discussion of the possibility of a 'progressive' carbon tax.
22. See E. von Weizsacker, *Earth Politics* (London: Zed Books, 1994), for an account of such arguments.
23. See discussion in T. Jackson, *Energy Efficiency Without Tears* (London: Friends of the Earth, 1992), pp. 40–1.
24. M. Pearson and S. Smith, *Taxation and Environmental Policy, Some Initial Evidence* (London: Institute for Fiscal Studies, 1990), p. 12.
25. This would still involve standards, but manufacturers who sold 'sub-efficiency standard' equipment would have to buy credits from manufacturers of equipment whose machines had 'above standard' energy efficiency. See M. Grubb, *Energy Policies and the Greenhouse Effect*, Vol. 1 (London: Dartmouth Press, 1990), p. 126.
26. M. Grubb, *The Greenhouse Effect: Negotiating Targets* (London: Royal Institute for International Affairs, 1989).
27. J.E. Macmahon, 'Quantifying Benefits and Costs of US Appliance Energy Performance Standards', in *International Energy Conference on Use of Efficiency Standards in Energy Policy* (Paris: IEA/OECD, 1992), pp. 15–21.
28. C. Gellings and J. Stuart McMenamin, 'DSM in the US: What's Real; What's Not and How Can We Tell the Difference', EPRI, Palo Alto, produced for WEC Conference, London, 1993.
29. S. Nadel, H. Geller and M. Ledbetter, *A Review of Electricity Conservation Programs for Developing Countries* (Washington DC: American Council for an Energy-Efficient Economy, 1991).
30. *ENDS Report No. 223*, August 1993, pp. 39–40. The fact that US states have freedom to set standards that are tighter than minimum federal requirements whereas individual EU states are stopped from setting higher standards than EU minimum levels is curious since the US is a single, if federal, country, while the EU is not a single country at all! The same story is true for motor vehicle emission standards.
31. E. Hirst, *Electric-Utility DSM-Program Costs and Effects: 1991 to 2001* (Oak Ridge, TN: Oak Ridge National Laboratory, 1993).
32. The highest proportionate unit price increases are around 2 per cent. This increased unit price includes payments for energy services and is not the same as increases in raw, primary energy prices. As was explained in Chapter 5, energy

efficiency programmes will tend to reduce the price of primary energy supplies.

33. Gellings and McMenamin, 'DSM in the US'.

34. The programme organised by the Dutch energy companies involves cuts in carbon dioxide emissions of 20 per cent below what they would otherwise have been in the context of a small reduction in the average consumer's energy bill. This compares with the UK's imposition of VAT on domestic fuel which will cut emissions by about 3 per cent in the domestic sector with a considerable increase in the average consumer's energy bill.

35. See D. Toke, 'A Dutch Lesson for Littlechild?', *Electrical Review*, 29 October 1993, pp. 32–5.

36. Energy efficiency is a low-cost means of alleviating fuel poverty, but it sometimes results in modest energy savings as the consumers may use the extra efficiency to heat themselves better rather than reduce consumption. The point is that energy efficiency services should be made available to all consumers.

37. Even in those areas of the US that have advanced DSM programmes, DSM measures will not normally be financed if they cost more than around 4 c/kWh.

38. See P.L. Joscow and D.B. Marron, 'What does Utility-subsidized Energy Efficiency Really Cost?', *Science*, Vol. 260, 16 April 1993, pp. 281 and 370; and several letters in response in *Science*, Vol. 261, 20 August, pp. 269–70.

39. Personal communication from F. Bisschop, Amsterdam Energy Company.

40. New power supply options that are available at any given time come in at a range of prices. In the UK, for example, the row about coal closures at the end of 1992 occasioned an investigation of the costs of bringing on stream new CCGTs. Analysis revealed that not only was there a considerable variation in the capital costs of new CCGT schemes, but also that the later proposals had to pay higher prices for natural gas. Cheap DSM measures would, in these circumstances, not only cut overall bills but also restrain unit price increases resulting from the purchase of relatively more expensive sources of new energy supply.

41. See discussion of such issues in Jackson, *Energy Efficiency Without Tears*, pp. 54–5.

42. Cambridge Econometrics, *UK Energy and the Environment* (Cambridge, 1994).

43. See, for example, R. Galli, 'Structural and Institutional Adjustments and the New Technological Cycles', *Futures*, October 1992, p. 779.
44. D. Mackenzie, 'World Unites to Fight Soaring Population', *New Scientist*, 23 April 1994, p. 5.
45. 'Population Misconceptions', *The Economist*, 28 May 1994, pp. 121–2.
46. O. Bernardini and R. Galli, 'Dematerialization: Long-term Trends in the Intensity of Use of Materials and Energy', *Futures*, May 1993, p. 433.
47. W. Walker, 'Information Technology and the Use of Energy', *Energy Policy*, October 1985, pp. 458–76.

8. Reducing Pollution from Motor Vehicles

1. S. Potter, 'The Need for a Systems Approach in Improving Transport Energy Efficiency in Europe', Proceedings of 1993 European Council for an Energy Efficient Economy (ECEEE) Summer Study, Vol. 2.
2. 'Road Transport, Pricing and Automation', *Energy Economist*, July 1993, p. 16.
3. World Wide Fund for Nature (WWF) report cited in *ENDS Report 217*, February 1993, p. 10.
4. A. Lovins, J. Barnett and L. Lovins, *Supercars, the Coming Light Vehicle Revolution* (Snowmass, CO: Rocky Mountain Institute, 1993).
5. Potter, 'The Need for a Systems Approach'.
6. L. Schipper, R. Steiner, M.J. Figueroa and K. Dolan, *Fuel Prices, Automobile Fuel Economy, and Fuel Use for Land Travel* (Berkeley, CA: Lawrence Berkeley Laboratory, 1993).
7. US information supplied by Lee Schipper in personal communication to the author; European information from P. Hughes, *Personal Travel and the Greenhouse Effect* (London: Earthscan, 1993).

9. Cleaner Coal?

1. Some analysts say that this process was artificially accelerated in the UK by the electricity and coal privatisation process.
2. E. Cassedy and P. Grossman, *Introduction to Energy* (Cambridge: CUP, 1990), p. 123.

3. An IGCC plant burning coal linked to a cogeneration system with good levels of overall energy efficiency will produce broadly the same carbon dioxide emissions as a CCGT.

10. Carbon Sequestration

1. 'Plant a Tree', *The Economist*, 24 October 1992, p. 92.
2. This calculation is made on the basis of figures supplied to me by R. Dewar, Institute of Terrestrial Ecology, Bush Estate, Midlothian, Scotland.
3. See costing assumptions made in D. Toke, 'Windpower and Sizewell B, a Cost Comparison', *SERA*, London, 1993.
4. F. Pearce: 'The High Cost of Carbon Dioxide', *New Scientist*, 17 July 1993, p. 27.

11. Nuclear Power

1. M. Damian, 'Nuclear Power, the Ambiguous Lessons of History', *Energy Policy*, July 1992, p. 600.
2. *Energy Economist*, December 1993, p. 32.
3. F. Krause et al., *The Cost of Nuclear Power in Western Europe*, Vol. 2, Part 3D of *Energy Policy in the Greenhouse* (El Cerrito, CA: International Project for Sustainable Energy Paths, 1994).
4. See, for example, A. Martensson, 'Inherently Safe Reactors', *Energy Policy*, July 1992, pp. 660–9.
5. Such as G. MacKerron, 'Nuclear Costs, Why Do They Keep Rising?', *Energy Policy*, July 1992, pp. 641–52; and F. Krause et al., *The Cost of Nuclear Power*.
6. MacKerron, 'Nuclear Costs'.
7. Krause et al., *The Cost of Nuclear Power*.
8. Chung-Taek Park, 'The Experience of Nuclear Power Development in the Republic of Korea', *Energy Policy*, August 1992, p. 730.
9. 1993 prices, 10 per cent discount rate, see Chapter 5 for some analysis.
10. Krause et al., *The Cost of Nuclear Power*.
11. *Wall Street Journal*, 25 January 1993, cited in Krause et al., *The Cost of Nuclear Power*.
12. J.G. Hewlett, 'The Operating Costs and Longevity of Nuclear Power Plants, Evidence from the USA', *Energy Policy*, July 1992, pp. 608–31.
13. *Nuclear Engineering International*, November 1993, p. 19.

14. There is a dispute about the extent to which these running costs represent avoided costs. British Nuclear Fuels (BNFL) claims few costs would be avoided because it has a 'binding' contract with Nuclear Electric to reprocess spent Magnox fuel. However, both BNFL and Nuclear Electric are state-owned so it is up to the government to decide whether the contracts are enforced.

15. E.F. Schumacher, *Small is Beautiful* (London: Vintage, 1993), p. 119.

16. A. Blower and D. Lowry, 'The Politics of Nuclear Waste Disposal', in J. Blundell and A. Reddish (eds), *Energy, Resources and Environment* (London: Hodder and Stoughton, 1991).

17. 'Fusion Test Heralds Clean Fuel', *Guardian*, 11 December 1993, p. 1.

18. See, for example, B. Keepin and G. Kats, *Greenhouse Warming: A Rationale for Nuclear Power?* (Snowmass, CO: Rocky Mountain Institute, 1989).

12. Renewable Energy

1. M. Slesser et al. (eds), *Macmillan Dictionary of Energy* (London: Macmillan, 1982), p. 237.

2. An informative and entertaining account of both grid-connected and off-grid wind turbines is provided by Paul Gipe in *Wind Power for Home and Business* (Post Mills, VT: Chelsea Green, 1993).

3. A. Cavallo, S. Hock and D. Smith, 'Wind Energy: Technology and Economics', in T.B. Johansson, H. Kelly, A.K.N. Reddy and R.H. Williams (eds), *Renewable Energy, Sources for Fuels and Electricity* (London: Earthscan, 1993), p. 123.

4. M. Flood, *Energy Without End* (London: Friends of the Earth, 1991), p. 20.

5. See research by Carl Weinberg, former R&D Director, Pacific Gas & Electric, cited in D. Bright 'The Role for Wind Will Grow', *Wind Power Monthly*, November 1993, pp. 20–1.

6. *Wind Power Monthly*, October 1993, pp. 16–17.

7. M. Brower, *Cool Energy* (Cambridge, MA: MIT Press,1992), p. 48.

8. See D. Olivier, *Energy Efficiency and Renewables, Recent Experience on Mainland Europe* (Credenhill, Herefordshire: Energy Advisory Associates, 1992).

9. I. Tzetzes, *Historiae*, ed. P. Leone (Naples: Libreria Scientifica Editrice, 1968), Ch. 2, lines 103–56.

10. Brower, *Cool Energy*, p. 51.

11. This should not be confused with photovoltaics, which, as we shall see, is concerned with the direct rather than indirect conversion of sunlight into electricity. Hence the use of the different terms 'solar thermal' and 'solar photovoltaics'.

12. 6 per cent discount rate, 1991 prices, project lifetime basis, P. Lequil, D. Kearney, M. Geyer and R. Diver, 'Solar Thermal Electric Technology', in Johansson et al. (eds), *Renewable Energy*, pp. 213–95 Please note that these projects would be much more expensive in places like the UK.

13. 6 per cent discount rate, P. Lequil et al., 'Solar Thermal Electric Technology', in Johansson et al. (eds), *Renewable Energy*, p. 281.

14. M.A. Green, 'Crystalline and Polycrystalline Silicon Solar Cells', in Johansson et al. (eds), *Renewable Energy*, pp. 337–60.

15. Gerald Foley, *Rural Electrification* (London: Panos Institute, 1989), p. 69.

16. PV costs have declined by a factor of ten since 1976 according to Brower, *Cool Energy*, pp. 61–2.

17. Foley, *Rural Electrification*, p. 70.

18. J. Josephon, 'Rural Electrification: Solar Can Prove It', *Environmental Science Technology*, Vol. 27, No. 7, 1993, pp. 1251–2.

19. J. Webb, 'Solar Power Brings a Warm Glow to Tyneside', *New Scientist*, 29 January 1994, p. 18.

20. Press releases from US DOE and USS Corp., January 1994, and personal communication from S. Guha (Vice President of USS Corp.).

21. 'The Light of the Sun She Loves', *The Economist*, 28 May 1994, p. 124.

22. D.O. Hall, F. Rosillo-Calle, R.H. Williams and J. Woods, 'Biomass for Energy: Supply Prospects', in Johansson et al. (eds), *Renewable Energy*.

23. Excluding energy used to process and transport the biofuels. This is usually not substantial, otherwise the fuel would generally be very expensive.

24. IEA, *The Role of Governments in Energy* (Paris, 1992); IEA/OECD, and US DOE, *Annual Energy Outlook*, 1993.

25. US Office of Technology Assessment study, 'Energy from Biological Processes' (Washington DC, 1980), cited by Hall et al., 'Biomass for Energy', in Johansson et al. (eds), *Renewable Energy*, p. 614.

26. Flood, *Energy Without End*, p. 28.

27. Aberdeen University Forestry Department, *Establishment and Monitoring of Large Scale Trials of Short Rotation Forestry for Energy*

(Harwell: UK Energy Technology Support Unit, 1989). 1988 prices have been converted to 1993 prices. The study assumed production of 12 dry tonnes per hectare per year.

28. Lynn Wright, *Commercialisation of Short Rotation Intensive Culture Tree Production in North America* (Oak Ridge, TN: Oakridge National Laboratory, 1989).

29. Hall et al., 'Biomass for Energy', in Johansson et al. (eds), *Renewable Energy*.

30. Patrick Knight, 'Sugar Alcohol for $32 a Barrel', *Energy Economist*, August 1993 pp. 10–12.

31. World Bank, *Energy Efficiency and Conservation in the Developing World*' (Washington DC: World Bank, 1993), p. 62.

32. L. David Ostlie, *The Whole Tree Burner* (Minneapolis, MS: published by the author, 1988).

33. *Electrical Review*, 31 July 1992, p. 21.

34. R.H. Williams and E.D. Larson, 'Advanced Gasification-based Biomass Power Generation', in Johansson et al. (eds), *Renewable Energy*, pp. 729–85.

35. J. Goldemberg, L.C. Monaco and I.C. Macedo, 'The Brazillian Fuel-alcohol program', in Johansson et al. (eds), *Renewable Energy*, pp. 841–63.

36. Patrick Knight, 'Sugar Alcohol for $32 a Barrel', *Energy Economist*, August 1993, pp. 10–12.

37. According to the US DOE, around 0.5 per cent of the US gasoline market was supplied by ethanol from corn wastes in 1990.

38. Miriam Hughesman, 'Ethanol from Waste?', *Energy Economist*, June 1992, pp. 8–10.

39. *NFFO News, Renewable Energy Review*, UK Department of Trade and Industry, 1993, pp. 16–17.

40. S. Meyers, N. Goldman, N. Martin and R. Freedman, 'Prospects for the Power Sector in Nine Developing Countries', *Energy Policy*, November 1993, p. 1126.

41. J. Dixon, L. Talbot and G. Le Moigne, *Dams and the Environment* (Washington DC: World Bank, 1989).

42. *The Economist*, 15 January 1994, pp. 50–1.

43. J. Cavanagh, J. Clarke and R. Price, 'Ocean Energy Systems', in Johansson et al. (eds), *Renewable Energy*, pp. 513–47.

44. T. Thorpe, *A Review of Wave Energy* (Harwell: ETSU, 1992).

45. Ibid.

46. Dave Ross, 'Osprey takes Flight at Dounreay', *Electrical Review*, 7 January 1994, p. 8.

47. Priscilla Ross, 'Groundswell of Enthusiasm for Geothermal', *Energy Economist*, September 1993, pp. 20–2.
48. J.Ogden and R. Williams, *Solar Hydrogen: Moving beyond Fossil Fuels'* (Washington DC: World Resources Institute, 1989).

13. A Low Cost Planet?

1. See, in particular, The Club of Rome, *Limits to Growth* (London: Pan, 1972).
2. World Commission on Environment and Development, *Our Common Future* (Oxford: OUP, 1987).

Index

Published by Pluto Press

TEARS OF THE CROCODILE
From Rio to Reality in the Developing World

Neil Middleton, Phil O'Keefe with Sam Mayo

❑ The first significant reassessment of the goals and objectives of the Rio summit

The industrialised world has turned its big guns on the poor. Threatened by a vast recession and a tottering financial system and struggling to extricate itself from the post-Cold War wreckage, it has put all its energies into self-defence by constructing fiercely protectionist trade blocs. Nowhere has this been more publicly apparent than at the much trumpeted UN 'Earth Summit', held in Rio de Janeiro in 1992.

Tears of the Crocodile examines what exactly happened in Rio, focusing specifically on the complex issues that perpetuate inequity, poverty and hunger. The authors offer a powerful argument that the Rio environmental agenda was about preserving Northern interests, and that these critical issues were not addressed.

ISBN hardback: 0 7453 0764 7 softback: 0 7453 0765 5

Order from your local bookseller or contact the publisher on
0181 348 2724.

Pluto Press
345 Archway Road, London N6 5AA

Published by Pluto Press

THE GREEN ECONOMY

Environment, Sustainable Development
and the Politics of the Future

Michael Jacobs

'A lucid and highly accessible account of how
industrialised economies can be redirected
to meet the environmental imperative.'
David Gee, Director, UK Friends of the Earth

'cogent and persuasive ... stands out on the
expanding shelves of green literature'
Financial Times

'refreshingly free of dogma ... it deserves serious
consideration by a wide readership'
New Statesman and Society

'Not the least of this book's merits is that it is extraordinarily
informative ... no one who reads it can fail to grasp the
central issues involved in creating an environmental
economics. For this reason, no less than for its vision and the
compelling logic of his arguments and practical proposals,
it represents a major step towards the reality of green
economics, sustainable development,
and undoubtedly the politics of the future.'
*George McRobie, co-founder of
Intermediate Technology Development Group*

ISBN hardback: 0 7453 0312 9 softback: 0 7453 0412 5

Order from your local bookseller or contact the publisher on
0181 348 2724.

Pluto Press
345 Archway Road, London N6 5AA

www.ingramcontent.com/pod-product-compliance
Lightning Source LLC
Chambersburg PA
CBHW021038210326
41598CB00016B/1062